Solid Waste Conversion To Energy

POLLUTION ENGINEERING AND TECHNOLOGY

A Series of Reference Books and Textbooks

EDITORS

RICHARD A. YOUNG

Editor, Pollution Engineering
Technical Publishing Company
Barrington, Illinois

PAUL N. CHEREMISINOFF

Associate Professor
of Environmental Engineering
New Jersey Institute of Technology
Newark, New Jersey

Additional volumes in preparation

Solid Waste Conversion To Energy

CURRENT EUROPEAN AND U.S. PRACTICE

HARVEY ALTER
Resources and Environmental Quality Division
Chamber of Commerce of the United States
Washington, D.C.

J.J. DUNN, JR.
Black & Veatch Consulting Engineers
Silver Spring, Maryland

MARCEL DEKKER, INC.   New York and Basel

Library of Congress Cataloging in Publication Data

Alter, Harvey.
    Solid waste conversion to energy.

    (Pollution engineering and technology ; v. 11)
    Includes index.
    1. Waste products as fuel. 2. Recycling (Waste,
etc.) I. Dunn, J. J., [date] joint author. II.
Title. III. Series.
TP360.A43   662'.8   80-14138
ISBN 0-8247-6917-1

MARCEL DEKKER, INC.

270 Madison Avenue, New York, New York 10016

Current printing (last digit):
10   9   8   7   6   5   4   3   2   1

PRINTED IN THE UNITED STATES OF AMERICA

PREFACE

Solid wastes have been the products and discards of man's activities since ancient civilizations. Before and during most of the industrial age there were cottage industries and institutionalized scavenging to recover products from household and industrial discards. For example, when a sailing ship was bought for scrap, everything possible was sold for reuse: timbers and decking were sold for furniture making, scaffolding, and so on; linen and canvas sails were cut into strips for papermaking, if not breeches; the tarred rigging ropes were unspun and turned into oakum to return to the sea; the untarred rope hawsers were respun to smaller ropes; and the leftover "shakings" were sold to make "cable papers." This was all done by hard, backbreaking work. With the Industrial Revolution came machines and new materials, which, while superior, could not be so easily recycled.*

Materials and energy became cheap, as did disposal. Recycling declined and waste accumulated. But the end of this era was coming. Threats of doomsday were often heard as part of the teachings of the new environmental awareness of the late 1960s. Overstatements emphasized the importance of the movement and its primary teachings. The halcyon days ended with the oil shortages of 1973-74.

Now there are concurrent and similar concerns among industrial countries worldwide about the need to properly dispose of wastes, while conserving materials and energy resources. A positive result of these concerns has been resource recovery, or the management and processing of solid wastes to recover usable products. The approach to this form of conservation is different in the United States and Europe due to different industrial practices, natural resources, economic bases, institutional histories, and so forth. However, the

---

* F. B. O. Harris, "Recovery and Re-use of Natural Fibers," Industrial Recovery 25(7) (1979).

approaches themselves, as well as the differences, can be instructive
for each to learn—and for each to teach—how to achieve common aims.
    The purpose of this brief book is to provide an introduction to the
background and technology of resource recovery from municipal solid
wastes, as practiced in the United States and Europe, including Great
Britain and Scandinavia. The idea is that while we do not have to copy
from one another, we can learn from one another's efforts. This is
not intended as a complete discourse on the subject. Some aspects—
such as source separation, materials policies, and materials recovery
practices and technologies in the United States—are omitted so as not
to repeat commonly available sources. Included, though, are the
similarities and differences of waste generation and composition,
mechanized materials recovery in Europe (including paper recovery,
something not practiced in the United States), technologies for mechanical processing for fuel recovery, and fuel use technologies.
    The descriptions are organized along generic methods of recovery,
rather than on individual projects and plants. For this reason, and
because of the difficulties in translating different monies and economic
systems to a common base, the costs of the various technologies are
not included. Rather, there is a description of a generalized way of
computing the likely cost of resource recovery in a community, especially at the early stages of planning when costs must be expressed as
imprecise estimates.
    Because there cannot be resource recovery without sale and reuse
of the recovered products, chapters are included that describe markets,
marketing, and specifications for recovered materials. Finally,
planning and risk management are briefly addressed. There are no
easy prescriptions here, and a longer discussion may merely be
sophistry.
    There is a tendency on both sides of the Atlantic to think, maybe
with a bit of romance, that the technology on the other side is "better."
An example is the thought among some in the United States that European incinerator technology (with or without heat recovery) is the
"proven" answer, if only because there exist a large number of repli-
cate plants. "Proven" should mean of predictable capital and operating
cost. This may be true in some locations, but it is noteworthy that
many Europeans come west to seek alternatives and that many are
developing systems for processing wastes to some form of refuse-
derived fuel. There is a misconception in the United States that incineration is widely practiced all over Europe. Compared to the United
States, it is, in terms of percentage of waste burned for disposal. It
may not be in terms of cumulative tonnage, simply because the smaller
countries in Europe have less to dispose of. However, incineration is

only widely practiced in such small countries as Denmark and Switzerland, which incinerate perhaps 60 percent of their municipal waste.
Other countries incinerate less: United Kingdom, 9 percent; Belgium
(Flanders), 29 percent; Germany (FRG), 22 percent; Netherlands, 25
percent; Italy, 33 percent; and Sweden, 33 percent. These land-short
countries—compared to the United States—do not have a history of
inexpensive disposal. Such facts, though available, have been buried
in obscure publications. It is hoped that this book will dispel some of
the old tales and put this practice, and some others, in proper per-
spective of current goals of resource conservation.

Resource recovery is an emerging technology, and thus plants
and practices change rapidly. Within this context we have tried to
make the text as currently correct as possible.

We direct and dedicate this book to the public and private sector
planners, managers, and designers who are trying to bring resource
recovery to their communities. Likewise, we direct it to the public
officials, industrial users, students, and teachers who will help them.
We hope our effort helps all.

Harvey Alter
J. J. Dunn, Jr.

CONTENTS

ACKNOWLEDGMENTS

This study was inspired by the Symposium on Solid Waste Conversion to Energy held in Hamburg, Germany, September 17, 1977 (see Appendix A). The meeting was sponsored by the International Federation of Municipal Engineers and International Solid Wastes and Public Cleansing Association, in cooperation with the International Union of Local Authorities and the Council for International Urban Liaison.

Many of the symposium papers are cited in the text and much of the information was used to develop concepts. The program and participants in the symposium presented are listed in Appendix A.

The study was further encouraged by many contacts with professional colleagues in the United States and Europe, all trying to learn from one another's experiences and problems. These contacts showed we can learn from one another. We need not merely copy to do so.

We thank the German Marshall Fund for their support of the writing of this text.

One of the authors, Dr. Harvey Alter, thanks the German Marshall Fund, NV Vuilafvoer Maatschappij VAM (Amsterdam), and the Municipal Environmental Research Laboratory, U.S. Environmental Protection Agency, for grants which supported trips to European plants in 1975, 1977, and 1978.

Special acknowledgment and thanks are due Dr. Michael C. Robinson, Director of Information, American Public Works Association, for his large contributions to Chapter 2 and for assistance in editing.

Solid Waste Conversion To Energy

Chapter 1

INTRODUCTION

As recently as the early 1970s, the industrialized nations enjoyed high
economic growth, supported in large measure by cheap energy. The
availability of petroleum products permitted economic expansion and
high employment as well as more leisure time and material comforts
than people had previously experienced. Then, for the first time in
modern history, aside from temporary disruptions during wars, these
countries were threatened by fuel shortages. This was a shock to
economies and life-styles so dependent on abundant, low-cost energy.

The realization that fuel supplies are finite, and the accompanying awareness that nuclear power was both environmentally and economically controversial, caused nations to accelerate the search for
new, alternative energy sources.

About a decade before the so-called energy crisis of 1973, there
was a new realization of the need for better management of solid wastes,
the discards of our society. Public works professionals and public
health officials were alarmed at the growing volume of wastes that
required disposal. In fact, it was less expensive to discard many items
than to repair or reuse them. The waste problem was, ironically, a
byproduct of affluence.

Societies have always generated waste. There is some evidence
that early civilizations picked up and moved on when their waste problem became overbearing. The ancient Greeks and Romans apparently
threw waste onto the floors of their houses and covered it with layers
of earth. When too much waste and earth accumulated, so that the
inhabitants could no longer fit through the door, they either raised the
roof or moved [2].

Early photographs of cities show streets littered with refuse and
animal manures. The history of solid waste management relates the
trials and successes of those who tried to dispose of wastes efficiently,
and in a safe and sanitary manner [3].

1

The period from 1968 to 1973 was a turning point in solid waste management. Inventors attempted new technologies for the disposal of solid waste, such as high-temperature or slagging incinerators to reduce the volume and destroy putrescible materials. They also recognized the need to control air emissions. Thus, some American cities began to abandon the traditional incinerator or crematory. Furthermore, greater emphasis was placed on using waste heat and recovering recyclable materials from refuse.

Resource recovery, or the systematic recovery and reuse of waste, involves the application of industrial technology to former practices. For centuries, scavengers gleaned materials from refuse, and by about 1900, picking lines were established in various parts of the world to institutionalize scavenging. The application of industrial technology required the substitution of capital for labor which resulted in large machines, facilities, and investments. Resource recovery held the promise of conserving material resources, reducing the quantity of material destined for disposal, and creating a revenue source to help defray disposal costs.

The new wave of resource recovery grew slowly until the energy crisis. Most solid waste is combustible; thus many cities in the United States used incinerators to "destruct" refuse. Some older installations used waste to generate steam and electricity, although few such plants were operating in 1973. At the same time, steam and/or electricity generation from solid wastes was commonplace in Europe. Teams of investigators traveled to Europe to inspect installations, and Europeans visited the United States to meet the visionaries developing new resource recovery systems [4]. There remains interest by both Americans and Europeans in one another's approaches to the recovery of energy and materials from solid wastes. This book is the product of of this mutual interest and willingness to cooperate.

ABOUT THIS BOOK

The objective of this study is to acquaint the reader with the materials recovery and waste-to-energy conversion and use technologies practiced in Europe. In doing so, comparisons are made to similar practices in the United States.

The comparisons are presented in a manner that allows the information to be incorporated into the municipal planning process for implementing such waste conversion methods. The benefits of conversion, in terms of conservation and lessened burden for disposal, are discussed.

Chapter 2 sets the stage for the treatment of waste-to-energy conversion by reviewing the recent history of electric power production
and consumption in the United States and Western Europe. Then follows
a computation, for both the United States and the European economic
community, of the fossil fuels which may be conserved by recovering
and reusing the secondary materials in municipal solid waste and by
using a portion of this waste as a fuel. The amount of energy conservation possible is sufficient to encourage further exploration of using
solid wastes for energy conversion and conservation.

Understandably, the amount of energy savings from waste depends
in part on the volume and composition of available waste. Such figures
are reviewed in Chapter 3 for the United States and some European
countries. The contrasting amounts and composition in various parts
of the world account, in part, for differences in waste management and
processing practices.

Chapter 4 reviews materials recovery processing technology in
Europe. A similar review for the United States is not included because
many other descriptions are available. Chapter 5 describes the technology of waste-to-fuel conversion and use. It is organized in a somewhat different manner than other studies on this subject. The technology of conversion is generically described by type of conversion. The
technology of refuse-derived fuel used to generate steam and/or electricity is organized along similar lines. This sets the stage for Chapter 6, which summarizes in tabular form details of many of the operating plants in Europe and North America.

The descriptions of technology ultimately should be used to assist
in the planning and implementation of resource recovery. Chapter 7
proposes one method of doing this. It provides a mechanism or tool
for computing the likely cost of waste-to-energy conversion and materials recovery to provide a basis for decision making. The decision
model is based on assured markets for recovered materials and energy products. There is no point in processing solid wastes for recovery
if the recovered products are not sold and/or used. Chapter 8 describes methods of marketing and securing firm commitments in
advance. The method should be applicable in both the United States
and Europe. Chapter 9 describes specifications for recovered products and Chapter 10 offers guidelines for further planning resource
recovery from municipal solid waste.

The presentation of European technology is purposely incomplete.
There is no discussion of composting municipal waste because so much
has already been written on this subject. It is clear that composting is
a mixed success story in Europe and has never been a major solid
waste management method in the United States.

There is no discussion of the recovery of materials from incinerator residues. Systems for processing such residues have been re-
searched and developed in the United States, Germany, France, England,
Denmark, Spain, and probably elsewhere since at least 1969. For a
variety of reasons, this technology has not been widely adopted except
for a few examples of ash or tailings being used for road stabilization
and a few scattered examples of steel recovery from residues. Perhaps this situation will change; the first residue processing plant was
to be commissioned in Holland in 1978. It is based on a process
developed in England by the Department of Industry, Warren Spring
Laboratory, and it has been rated to treat 25 tons of clinker per hour
to recover copper, aluminum, ferrous metals, and "sand" for road
building.

There are no discussions of pyrolysis or anaerobic digestion of
solid wastes. These technologies are just starting and there are few
examples beyond the research stage. For example, a recent survey
of pyrolysis, thermal gasification, and liquefaction of solid waste and
residues reported six such systems in North America and two in
Western Europe operating on refuse and possibly sludge [5]. Only two
in the United States may be considered commercial or operating units.
At least one listed in Europe is an operating unit.

The emphasis of this book is clearly on municipal solid waste,
particularly the household portion, since worldwide developments
focus on this portion of the waste. This book is not a complete description of resource recovery. Rather, it is a limited discussion of
important aspects of the technologies and practices on two continents.
The discussion should acquaint readers with the state of the art and
how and why it differs in Europe from the United States. Hopefully,
the study will provide a point of departure for further advances.

Wherever possible, references to other literature are given.
Appendix B lists addresses for contacting many of the European organizations. Note that all quantities are expressed in the International
System of Units, or in metric tons (or tonnes), unless expressly noted
otherwise.

In describing European technology and practices, the term
European is used in the broad sense to include Scandinavian countries
and Great Britain. This is done as a convenience, and the authors
apologize for taking such license.

Finally, there is much in this field the United States can learn
from Europe and vice versa. Workers on both sides of the Atlantic
need not and should not just copy one another. Countries have different problems. Hence, they should have different practices and different solutions.

NOTES AND REFERENCES

1. The same is true about effluents to the air and water, but these wastes are well outside the scope of this book.
2. C. G. Gunnerson. Debris accumulation in ancient and modern cities. J. Eviron. Engr. Div., ASCE 99: 229-43 (1973).
3. E. L. Armstrong, M. C. Robinson, and S. Hoy (eds). History of public works in the United States, 1776-1976. Amer. Public Works Assoc., Chicago, 1976.
4. This study focuses on developments in the United States and Europe. It was not the authors' intent to overlook or minimize developments in other parts of the world, particularly Canada and Japan. Some examples of Canadian contributions are provided.
5. J. J. Jones, R. C. Phillips, S. Takoaka, and F. M. Lewis. Pyrolysis, thermal gasification, and liquefaction of solid wastes and residues: Worldwide status of processes as of fall, 1977. In Proceedings, National Solid Waste Processing Conference. ASME, New York, 1978.

Chapter 2

ENERGY CONSUMPTION HABITS IN THE
UNITED STATES AND EUROPE

In comparing waste-to-energy conversion in the United States and
Europe, it is important to understand differences in power generation
and consumption habits. The following discussion traces energy consumption and electric power generation to illustrate the growth and
changes in different countries.

Solid waste was first used as a fuel to generate steam just before
the turn of the century in both the United States and Europe. Since
about 1945, the use of waste for steam and electricity generation has
grown in Europe. During most of the same period, the practice was
not widely implemented in the United States. But since about 1973,
greater attention has been focused on the feasibility of burning processed solid waste (refuse-derived fuel) in utility power plants — either
by itself or as a supplement to coal — to generate process steam
and/or electricity. Some plants are operating in the United States,
and this approach is being followed closely in Europe.

Industrialization and rising living standards are naturally accompanied by a growth of power generating capacity. Since 1940 the
United States' electric-generating capacity has doubled every decade,
and from 1940 to 1970 the per caput consumption increased more than
sixfold. The United States has consistently held first place among the
world's nations in the production and use of electricity.

GENERATION AND CONSUMPTION IN
THE UNITED STATES

The total installed generating capacity in the United States in 1974 was
nearly 460 gigawatts (10^9 watts), more than twice the installed capacity of the Soviet Union, the second largest producer. Since 1974, there
has been a marked slowdown in the growth of electrical consumption,
largely growing out of reduced supplies and higher prices for fuel.

6

However, as population increases, total consumption will probably increase, though at a lower rate than before [1].

In 1974 some 80.5 percent of all electrical generation in the
United States was by conventional steam plants; 14.7 percent by hydroelectric, including pumped storage; 4.5 percent by nuclear; and 0.3
percent by internal combustion. Almost 54 percent of this power was
produced from coal, 21 percent from gas, 20 percent from oil, and 5
percent from nuclear fuel [1].

In the United States, the electrical utility industry includes more
than 3600 systems. These are usually divided into four different
types: investor-owned companies, nonfederal public agencies, federal
agencies, and rural electric cooperatives. The investor-owned systems operate nearly 80 percent of the generating capacity and sell
more than three-fourths of electric power consumed. The 200 largest
investor-owned companies account for more than three-fourths of the
nation's generating capacity [2].

Nonfederal public agencies, including municipal, county, state,
public utility districts, irrigation districts, and similar systems account for 9 percent of the country's electric energy capacity and about
13 percent of electric energy sales. About two-thirds of the 2075 nonfederal electric systems purchase all of their power from other systems. Approximately one-third generate part of their requirements,
and some sell considerable quantities of power. Most of the nonfederal
systems were created by municipalities to serve customers within
their jurisdictions. Competition between municipal systems and
investor-owned companies occurs infrequently, and in many instances
public and private electrical utilities are interconnected.

Federal electric systems make a substantial contribution to the
utility industry. Nearly 10 percent of the total electrical generating
capacity was federally owned in 1976. The Bureau of Reclamation,
the Corps of Engineers, and the Tennessee Valley Authority are the
agencies primarily responsible for the construction, operation, and
maintenance of federal power projects. Several small projects en-
compassing electric power facilities have been constructed by the
Bureau of Indian Affairs and other federal agencies. At mid-1974,
there were 77 federally owned hydroelectric plants — 67 in operation
and 10 under construction.

The electric power industry began in the United States following
Thomas A. Edison's invention of the first commercially practical
incandescent lamp in 1879. Early systems, like the one he inaugurated
in New York City in 1882, had the generator located near the area of
consumption and used direct current. The nation's first hydroelectric
station, at Appleton, Wisconsin, began operations in 1882. A further
step forward was made four years later with development of an alter-

nating current system that facilitated long-distance transmission of
electricity. Before the turn of the century, the electric power industry evolved into three distinct segments: generation, or power production; transmission, or long-distance transportation of electrical
energy; and distribution, or delivery of the electricity in the area
where it was consumed. The demand for electricity, resulting from
rapid acceptance of electric lighting and improvements in the electric
motor for street railways and industrial use, led to development of
utility systems capable of providing more than a gigawatt of electric
generating capacity by 1900.

In the early years of the twentieth century, until 1917, supply systems developed rapidly to meet the ever-expanding demand. Agreement on a 60-hertz frequency in the United States in 1917 was an
important milestone in standardization. Techniques of electricity
transmission were progressively improved, and more electric supply
systems were in operation in 1917 than before or since. The 1920s
witnessed accelerated expansion of electric power companies as part
of the period's business prosperity. Efficiency was increased considerably by larger generating units and improved power plant design.
For example, the average amount of coal required per kilowatt-hour
was reduced from 3 pounds (1.4 kg) in 1920 to 1.62 pounds (0.73 kg) in
1930. The proportion of the population served by electric utilities rose
from 35 to 68 percent during the decade.

The 1920s witnessed the greatest growth the nation has experienced in electrification. Economies of scale and generation, with con-
comitant increases in capital requirements, fostered consolidation.
The holding company became the means in the 1920s to keep pace
with the burgeoning growth in electric demand. It could command
needed capital and marshal engineering talent and management know-
how. By 1929, some 60 percent of the nation's total generating capacity was controlled by seven holding companies. Through the 1920s,
until the mid-1930s, investor-owned systems accounted for more than
90 percent of utility generating capacity. But the growth of municipal
systems, federally owned power projects, and rural electric cooperatives in subsequent years reduced the proportion served by the private
sector.

The economic expansion from World War II to the early 1970s was
beneficial to the electric utility industry and for the maintenance of a
high rate of return for the industry. During this period, the industry
experienced sustained economic growth. The expansion of the electrical power industry was an integral part of the overall growth of the
American economy. It was not difficult for electric utility managers
to grasp and act upon the main technological, economic, and regulatory
factors and trends. The technological foundations of the industry were

well established, and the options were limited and well defined. For steam-electric generation, the choice of fuels was primarily coal and natural gas. Fuel oil was used only marginally. Generally, fuel was relatively cheap and a buyer's market prevailed. Through techno-logical innovations, the amount of steam required to generate a kilo-watt of electricity steadily declined. Also, the capacity of base-load steam generators increased during the period from 200 to 1000 mega-watts.

With steadily increasing demand for electricity, the utilities gen-erally found it advantageous to add new, larger, and more efficient generators. Overall, the price of electricity tended to fall relative to the consumer price index until the late 1960s. This was undoubtedly a factor in stimulating the consumption of electricity. The more than two decades of steady and rather predictable progress came to an end with the onset of several problems in the late 1960s. Increases in generation economy became more difficult because the size of indi-vidual turbine generators reached the maximum limit imposed by metallurgical factors. The adoption and operation of commercial plants placed unprecedented demands on the engineering and mana-gerial skills of electric utilities. Public concern over the effects of air and water pollution and the subsequent enactment of laws to im-prove and protect the environment imposed constraints on the industry and slowed the transition to nuclear power generation. Costly switches to low-sulfur fuels in existing fossil power plants were required to meet air quality standards, and expensive emission control devices drove up generating costs.

The past abundance of low-cost fuels for power generation de-clined as natural gas supplies fell in the late 1960s. The dramatic increase in oil prices following the oil embargo in 1973 resulted in a shift from a buyer's to a seller's market in fossil fuels. In the five years following 1969, the price of electricity rose as fast as, but no faster than, the consumer price index. In 1974, after 15 years in which demand grew at an average rate of 7.4 percent, there was a drop to a growth rate of slightly more than 1 percent. The rate of growth and demand recovered in 1975 from the 1974 recession-year level, but did not reach the 7.4 percent pre-1973 level. Events follow-ing the inception of the fuel crisis in late 1973 seriously undermined the economic health of the utilities industry and dampened investor confidence. Many companies experienced substantial declines in earnings, and some even faced bankruptcy.

The search for new and improved energy sources for producing electricity dramatically increased following the energy crisis. The jump in fuel costs helped spur research and development. In April 1974, average prices paid by utilities for coal, oil, and gas were respectively 63, 170, and 31 percent higher than one year earlier. Large-scale research and development programs to alleviate the energy crunch began and included a broad panorama of activities. The efforts to help resolve the fuel problem for the electric utilities were directed toward making coal (the most abundant and available fuel) more environmentally acceptable, further developing nuclear technology, and finding means to utilize new energy sources.

Coal deposits are estimated to be 3×10^{12} U.S. tons and account for almost 80 percent of the United States' fossil fuel reserves. At current consumption rates, coal reserves would provide supplies for more than 500 years. The production rate of this resource, which represents about 17 percent of United States energy consumption, did not increase appreciably from 1943 to 1975. In the late 1960s, however, there was a shift from the use of coal for electrical generation to oil and gas because the latter two fuels were still relatively cheap and more environmentally acceptable. Burning coal posed environmental problems of emissions of particulates and sulfur oxides [3].

At the beginning of 1976, there was adequate capacity to provide the nation with the electricity demanded. There was also a wide consensus that in the future Americans would be expecting electric utilities to provide a greater share of the nation's energy. With the shortages of domestic oil and natural gas, it seemed logical to shift the energy base from oil and gas to coal and nuclear energy. These two sources, in turn, are most usable in the form of electricity.

GENERATION AND CONSUMPTION IN
EUROPE

The high rate of increase in the United States' electrical energy demand during the first half of the twentieth century was paralleled by a rise in the installed capacity of generating plants in Europe. At first, an apparent abundance of coal and undeveloped water power in the European industrialized countries enabled production to advance without concern for limited resources. The discovery of large reserves of petroleum had the same effect. However, as the rate of use of these resources in relation to the total proven reserves rose rapidly, the position became disquieting. In the decade following World War II, the electrical supply industry was greatly concerned with the introduction of new sources of energy.

While many sources were studied intensively, it became clear that nuclear energy offered the greatest promise as a major future source. However, the discovery of huge resources of petroleum, especially in the Middle East in the late 1950s, changed the energy situation. In the

1960s, petroleum proved to be an abundant and cheap fuel, while hydro-electric resources neared exhaustion and domestic coal in many developed countries increased in cost. Thus, many developed countries imported large amounts of Middle East oil for power generation. Development of nuclear power was delayed by this cheap oil. Europe, in time, came to rely on imported oil for some 40 percent of its total energy, and the rate of dependence reached 70 percent in 1970. After a decade of cheap oil, its price rose sharply in the early 1970s because of embargoes, quota restrictions, and price increases [4].

In Great Britain, by 1925, 572 organizations provided electricity and 491 generating stations were operating with total capacity of 4.4 gigawatts. World War I demonstrated the need for national planning, and the 1926 Electricity Act established the Central Electricity Board to be responsible for transmission facilities and interconnection of selected generating stations. The construction of a 132-kilovolt national grid, standardized at 50 hertz, was completed by 1938. After World War II, the shortage of generating plants and the need for reorganization were apparent. For several years, load shedding caused blackouts and brownouts, and the discouragement of energy consumption was common. Such steps did not become unnecessary until 1956.

In April 1948, the British electric industry was nationalized and about 95 percent of the generation came under government control. In 1957, the Central Electricity Generating Board, responsible for both electricity generation and transmission, was established. Twelve area boards for electricity distribution were also formed. The number of generating stations was reduced, and new ones were built under a succession of annual construction programs that dramatically increased the country's installed capacity. Great Britain's nuclear power generation program was the largest of any country in the world by the early 1970s. After a 10-year program, the total capacity of eight nuclear power stations was 4.7 gigawatts, and the electricity production was 28.5×10^9 kilowatt-hours in 1972. These nuclear power installations accounted for 12 percent of the total production in the United Kingdom and about 3.3 percent of the total nuclear power generation in the world. Another special feature of the United Kingdom's utility industry is the low load factor of the system. The yearly load factor lies between 60 and 70 percent in many countries, but in Great Britain it remained about 45 percent for long periods, and only in the 1960s rose to 55 percent. To increase off-peak demand, use of electric room heating and electric water heaters was being encouraged by introducing special rates [5].

Before World War II, the generation and distribution of electricity in France was entirely private except for a small number of municipal companies and the Companie Nationale du Rhone, which was a mixed

public and private utility. Generation was about half by steam and half
by hydro, with the latter being located in the southern part of the country. The general movement for nationalization following the end of
World War II was the direct cause for the decision to nationalize the
French power industry in 1948. Certain other factors were also involved. The excessive number of small companies resulted in a considerable variety of rates, poor coordination, and unreliable service.
Vast expenditures to reestablish a system that had suffered from the
war and an expected rise in demand were also compelling factors. In
the past thirty years, there has been a progressive decline in the
fraction of hydroelectric capacity as a part of the total energy resources. This falling off was a result of the exploitation of remaining
underdeveloped hydroelectric sites and the poor economics of hydroelectric construction and hydrogeneration compared to steam. Still
another change was the marked increase in the size of steam-electric
units [5].

West Germany had retained a substantially private power industry.
In the early 1970s, there were approximately 4000 utilities and nearly
4100 industrial installations generating or distributing some or all of
their own electricity in West Germany. Though most of the electric
utilities were private enterprises, most were partly financed by public
organizations; purely private industry generated only 4 percent of the
total electric power consumed. Industrial plants contributed a large
share, more than 30 percent of the total power production. Part of
this industrial power is fed into the utility network, accounting for as
much as 50 percent of total output. West Germany relies heavily on
interconnections with other countries to meet its electric energy demands. Suitable sites for pumped storage and other hydroelectric
stations are scarce, and some serving West Germany are located in
other countries, such as Luxembourg [5].

In Italy, before nationalization of its electric power industry in
1963, nearly 75 percent of the total electricity was produced by private utilities. With nationalization, a single government organization
was established to generate and sell almost all of the electricity in
the country. Although hydroelectric capacity has remained essentially
constant, its share of total capacity has fallen from 82 percent in 1960
to less than 36 percent by 1970. Oil provided the main fuel for this
change, and its consumption increased almost 10 times during the
1960s. Italy produces sizeable amounts of geothermal power. The
first geothermal production in the world began there in 1905. The
capacity of the early 1970s was nearly 0.4 gigawatt, and yearly production reached some 3 percent of the total electricity produced [5].

FUTURE CONSUMPTION

It is not easy to predict future demand patterns for electrical energy. Even though past trends are quite regular, contemporary forecasts are complicated by unclear outlook of national energy policies and profound environmental obstacles — real and imagined — to the construction of nuclear power plants and other generation facilities. In some respects, however, the general outlook for electricity is easier to determine than predicting overall energy requirements.

From 1920 to 1972, world generation of electricity increased at an average growth rate of 7.6 percent, corresponding to a doubling of production every 9.5 years. From 1952 to 1972, the development was even faster, with an average yearly growth rate of 8.2 percent resulting in a doubling every 8.8 years. During the same two decades, in the nine countries comprising the European Economic Community (EEC), electric power demand rose 7.5 percent per year, constituting a doubling every 9.7 years. The corresponding figures for the United States are similar: 7.2 percent per year, a doubling every 10 years [6].

It is quite possible that a predicted decrease of the total energy growth rate will be accompanied by an increase in consumption of electricity. The most important reason is that in terms of use, electricity is a clean and convenient form of energy. Many future scenarios view electricity playing a greater role in automated industries, public transportation, household space heating, and agriculture. Electric power from coal or nuclear plants may replace space heating and industrial power production now provided by oil or gas.

The importance of electricity within the energy sector on both continents appears likely to continue in the foreseeable future. In 1970, the share of primary resources used to produce electricity was 25 percent in the United States and 23 percent in Europe. By 1985, the respective figures are likely to be on the order of 37 and 34 percent [7].

Though electricity consumption has followed a sustained (and perhaps predictable) growth rate, total energy consumption patterns changed radically in the 1970s. The principal feature of the current energy embroglio is that the United States and Western Europe have become heavily dependent on imported oil. This has been a gradual process that has taken place over the past quarter-century. Despite a steady rise in demand for crude oil in the 1950s and 1960s, an oversupply persisted. The discovery of enormous reserves in the Middle East, followed by significant finds elsewhere, glutted the world market. Oil-exporting nations found markets for their commodity in Western Europe; Japan's imports increased gradually; and the United States

remained practically self-sufficient. Therefore, Europe was the scene
of fierce energy competition. Demands for gasoline and other petroleum products grew rapidly, and fuel oil, offered at very low prices,
successfully competed with coal and retarded its production. In the
United Kingdom, for example, the price of industrial coal soared upward by more than 60 percent from 1955 to 1970, whereas the price of
industrial fuel oil rose by just over 3 percent during the same period.
The price advantage of oil was greatly enhanced by the convenience
and lower relative transportation costs of liquid over solid fuel. The
result was that oil rapidly became the most important primary source
of energy in Western Europe [4].

In Europe, the shift away from coal meant that the area, which in
1960 still met two-thirds of its energy needs with domestic production, in 1970 produced not more than about 40 percent of its needs.
Figures varied, of course, from country to country because of the un-
even geographic distribution of domestic resources. Denmark, for
example, has almost no indigenous energy resources, whereas the
Netherlands produced some 70 percent of its consumption. Domestic
production in 1970, expressed as a percentage of total consumption,
ranged from 18 percent in Italy to 31 percent in France and 55 percent in both West Germany and Great Britain. European coal production, once the lifeblood of the Industrial Revolution, contracted significantly. By the early 1970s, only Great Britain and West Germany
mined hard coal on a significant scale, and the once-considerable
French and Belgian production had dropped sharply. Oil had become
the most important source of energy in every Western European
country and the United States by 1970.

In 1960 the energy self-sufficiency of the six major European
countries was about 60 percent; it fell to 40 percent by 1973. This
trend bore a close relationship to the declining use of coal as a primary fuel. In 1960, coal accounted for 64 percent of the total, in 1973
only 24 percent. Oil, on the other hand, grew from 28 to 58 percent
during this period. Natural gas gradually filled the gap; other sources
(nuclear, hydro, and geothermal) remained fairly constant [4].

EVOLVING ENERGY POLICIES

To insure reliable energy supplies and protect national economic sta-
bility, North American and Western European countries have adopted
a variety of energy policies. Central governments have used taxation,
regulation, subsidies, and tariffs to foster a desired energy consumption pattern.

Throughout the 1960s, governments of large European coal-producing countries subsidized coal production and taxed oil consumption to abate the shift from coal to oil and prevent widespread dislocations in the mining labor force. In the Netherlands, however, the discovery of huge gas reserves in the mid-1960s prompted the government to encourage the transition to gas by imposing high taxes on oil and granting financial assistance to public gas utilities.

Aside from fiscal policies designed to influence the energy mix, some countries have tightened centralized governmental control of the energy market. France and Italy promoted national oil companies against the competition of foreign multinational corporations to assure the availability of petroleum supplies. The French government divided the domestic petroleum market among major oil companies and regulated prices and imports of oil products. The Dutch government owns a large share of the natural gas industry. West Germany has promoted an oil consortium in which it has a majority share. In Britain the coal industry is nationalized as is the distribution of gas. The oil market has been traditionally open to multinational corporations, but the government is now seeking acquisition of majority shares in the North Sea operations of the major oil companies. The British government requires all North Sea oil and gas to be delivered in the United Kingdom so that it can better control the domestic and foreign distribution of the resources. In the United States, measures have been undertaken to increase domestic production, promote conservation, and foster research and development within the framework of existing tax and regulatory legislation [8].

Throughout the Western world, countries are tightening their control over energy resources and attempting to insure that energy and environmental policies are in accord. Historically, an increase in gross national product for European and North American nations has generally been paralleled by an increase in energy consumption. However, the rise in energy prices, proposed conservation practices, and environmentalist opposition to some types of energy development may significantly change this relationship. Some growth critics argue that rises in GNP are not necessarily related to the quality of life and human welfare. Nevertheless, a recent report points out that "adequate energy supply is the only means by which the physical energies of human beings can be multiplied so as to produce the food, goods and services which are valued for their own sake, but are also the basis of the divisions of labor that have made possible the rapid and largely beneficial growth of social services, including health care and education" [9].

The energy upheavals of the mid-1970s compelled North American and Western European nations to give greater attention to the funda-

mental role of energy as the price of goods soared, some commodities
became scarce, production in many industries fell off, and workers
were idled.

The crisis marked a historical turning point. Throughout the
twentieth century, economic growth and industrial development have
always been paralleled by increased energy consumption. In fact,
during the twentieth century no Western country has been able to sustain increases in its gross national product without the expenditures
of additional increments of energy. There were a few years in various
countries where significant income gains were accompanied by small
increases or even decreases in energy consumption, but these rarely
occurred and generally marked the onset of general economic down-
turns. Furthermore, small reductions in energy consumption have
taken place where there was a corresponding rise in energy use efficiency (such as in the 1920s), but such instances are rare.

In the final analysis, gains in real personal per caput income and
per caput total energy and electricity usage have fundamentally followed
consistent and parallel courses. North America and Western Europe,
therefore, are currently facing an unprecedented dilemma: how to
maintain or elevate current standards of living in the face of an ever-
growing host of factors that raise the prices and restrict the availability of traditional fuels.

Visionaries who forecast and espouse the attributes of the so-
called postindustrial age contend that the developed nations will learn
to reduce energy consumption without significantly reducing the prevailing quality of life. If, however, standards of living and gross
national product are inextricably linked to energy consumption, then a
means must be discovered of overturning what seems to be a historical
imperative. Thus far, no country has discovered how to reduce energy
consumption in the long run and at the same time increase or maintain
national prosperity. Furthermore, the past suggests that prolonged
economic stagnation often causes domestic political and social upheavals as well as international tensions that threaten world peace.
Oil, for example, has become an instrument of economic warfare.
Therefore, for the oil-importing countries, the availability of energy
resources has become a fundamental problem of national economic se-
curity. The record suggest that if the Western developed nations are to
maintain their well-being, which has been based on inexpensive energy,
they must face up to the foregoing historic economic realities [9].

TAPPING NEW ENERGY SOURCES

Since 1973, Western Europe and the United States have sought means
to reduce oil imports. Fundamentally, three courses of action seemed

open to them. First, they could undertake energy policies that focused on reducing waste and increasing the efficiency of primary energy production, conservation, and use. Second, they could adopt programs for accelerating the production of indigenous resources (mainly coal, nuclear power, and offshore oil and gas). Third, they could develop alternate sources of energy, including the use of solid waste as fuel.

Emphasis is being placed on beginning a transition away from oil as a primary energy source. This means developing various sources of other fuels, such as geothermal and solar energy and the utilization of wastes. The other side of the coin is conservation. Nonuse is an unrealistic alternative.

ENERGY CONSERVATION AND FUEL PRODUCTION
BY PROCESSING SOLID WASTES

The municipal solid wastes of the United States and Europe contain recoverable metals, glass, and paper — useful as secondary materials for the manufacture of "new" materials — and a mixture of combustible materials useful as a fuel or as a feedstock for conversion to some other form of fuel.

From the standpoint of total energy expenditures, using secondary materials in manufacturing requires less energy than if virgin materials are employed. The energy required to convert iron ore, coal, and wood fiber to more refined and usable materials is saved. The amount of energy which can be conserved by using secondary materials recovered from municipal solid waste, and the amount of energy which can be generated using municipal solid waste as a fuel, have been calculated and reported [10-13]. Most such analyses compute the higher heating value of the combustible portion of the waste (namely, the heat of combustion at constant volume) for the current and prospective quantities of waste generated. To this scenario are often added several others, such as the effect and likelihood of certain changes in the composition of the waste stream. While these are seemingly valid approaches, they must rely on interdependent assumptions about the future.

A perhaps more conservative approach is reviewed here. In this approach, the energy potentially recoverable or conservable from the municipal solid waste generated in the United States in 1971 and 1974 is computed. This is done by first estimating the amounts of recoverable materials in the waste stream and using this result to further estimate the fossil fuels consumption which would be averted by use of the materials and by use of the combustible portion as a fuel. The results of this analysis are summarized here as a means of expressing

solid waste as an "energy source." The full details of the analysis
have been reported [14].

An estimate of the amounts of materials potentially recoverable
from municipal solid waste in the United States Standard Metropolitan
Statistical Areas (SMSAs) is computed in Table 2.1. An estimate of
the energy saving potentially available from using these recovered
materials is computed in Table 2.2. Table 2.3 shows similar results
from some of the materials potentially recoverable from the waste in
the European Economic Community (EEC), as a comparison. Table
2.3 was developed using published data for the amounts of materials
discarded in the EEC [15] and the energy savings and recovery efficiencies in Tables 2.1 and 2.2. The total of Table 2.3, for the EEC, is
about 2×10^{17} joules (J) per year (1.9×10^{14} Btu/y), about half that
for the United States (Table 2.2).

The total savings listed in Table 2.2 may be compared to the total
energy consumption in the United States in 1973, which was 7.88×10^{19}
J. The potential savings of energy from materials recovery and reuse
is thus on the order of 0.5 percent of total energy consumption.

TABLE 2.1. Amounts of Recoverable Materials in the U.S. Municipal
Waste Stream (SMSAs only)

Material	Weight % in waste	Amount available,[a] SMSAs, 1971 (Mtons)	Amount available,[a] SMSAs, 1974 (Mtons)	Estimated recovery efficiency %	Estimated potential recovery (Mtons/y)
Magnetic metals	7.0	6.10	6.80	85.0	5.8
Aluminum	0.7	0.61	0.68	65.0	0.36
Other non-ferrous metals	0.3	0.26	0.29	75.0	0.18
Glass	9.0	7.86	8.74	64.0	5.0

[a] Based on waste generation of 87.1 Mtons in 1971 and 97.1 Mtons in
1977, SMSAs only.
Source: Ref. 14.

TABLE 2.2. Computed Energy Savings Which Would Have
Been Realized if Amounts of Materials Listed Had Been
Recovered and Reused, United States (SMSAs only)

Material	Potential recovery (Mtons/y)	Energy savings (GJ/ton[a])	Energy savings (10^{17} J/y)
Magnetic metals	5.8	49.0	2.84
Aluminum	0.36	281.0	1.01
Other nonferrous metals	0.18	74.7	0.13
Glass	5.0	2.91	0.15
		Total:	4.13

[a]$GJ = 10^9$ joules
Source: Ref. 14.

TABLE 2.3. Computed Energy Savings Which Would Have
Been Realized if Amounts of Materials Listed Had Been
Recovered and Reused, EEC

Material	Potential recovery (Mtons/y)	Energy savings (GJ/ton[a])	Energy savings (10^{17} J/y)
Iron and steel	2.52	49.0	1.24
Aluminum	0.21	281.0	0.58
Glass	5.33	2.91	0.16
		Total:	1.98

[a]$GJ = 10^9$ joules.

The magnitude of the contribution of the combustible portion of municipal solid waste as a new energy or fuel source was estimated in a similar manner. The result was a potential savings of 9.11×10^{17} J/y (8.6×10^{14} Btu/y). Depending on how this combustible portion is used as a fuel, i.e., depending on the thermal efficiency of use, this figure may be adjusted slightly upward or downward. (Details are given in Ref. 14.) The corresponding figure for the EEC is of the order of from 3 to 5×10^{17} J/y [16]. The potential savings from using the combustible portion of municipal solid wastes in the United States (SMSAs only) is then of the order of 1.2 percent of the total United States energy consumption in 1973.

Although this may seem like a small percentage, two observations must be made. The first is that the energy savings from municipal solid waste available (United States, SMSAs), from both materials and fuel recovery, is the arithmetic equivalent of approximately 10^6 barrels of oil per day (1.6×10^5 m^3 or 160×10^6 liters) and even more if some other wastes, such as from industrial or agricultural sources, were processed. (Arithmetic equivalence is defined in Chapter 5. It is the usual way of expressing such equivalencies.) The second observation is that the waste in the urban areas is where it is needed as a fuel. The analogy of solid waste as an "urban ore" has been used to the point of almost being trite. Nonetheless, it may well be timely to extend the analogy to waste as an urban oil well or coal mine.

SUMMARY

The history of energy use (at least in the form of electricity) in the United States and Western Europe shows recent decades of large growth and a more recent period of concern and conservation. The latter was brought about by the cartel action of quickly restricting supplies and raising oil prices. It was also brought about by the realization that fossil fuel reserves are finite. That is, they can and will someday be exhausted.

One of the many strategies at hand for conserving fossil fuel supplies is the use of secondary materials recovered from municipal solid waste. Another strategy is utilization of the combustible portion of solid waste as an alternative fuel.

NOTES AND REFERENCES

1. Edison Electric Institute. Statistical yearbook of the electric utility industry. New York, 1975.

2. The following overview of the electric power industry is largely based on E. T. White. Utilities. U.S. Government Printing Office, Washington, D.C., 1976; J. Bauer and P. Costello. Public organization of electric power. Harper, New York, 1949; Edison Electric Institute. Historical statistics of the electric utility industry through 1970. New York, 1973; National Economic Research Associates. The development and structure of the electric utility industry and the impact of government policies. In U.S. Congress, Senate, Committee on Interior and Insular Affairs, 93rd Cong., 2nd sess., Electric utility policy issue. U.S. Government Printing Office, Washington, D.C., 1974; U.S. Library of Congress, Congressional Research Service. The electric utility sector: Concepts, practices, and problems. Prepared for the subcommittee on energy and power of the Committee on Interstate and Foreign Commerce, House of Representatives. U.S. Government Printing Office, Washington, D.C., 1977.

3. White. Utilities. (op cit.)

4. F. G. Ray. Western Europe and the energy crisis. Trade Policy Research Centre, London, 1975.

5. Electric power. Encyclopedia Britannica, 8: 616–37 (1976); P. Spron. The social organization of electric power in modern societies. (MIT Press, Cambridge, 1971; United Nations, Economic Commission for Europe. Developments in the situation of Europe's electric power supply industry during the post-war period. United Nations, Geneva, 1957.

6. T. Leardini. Electric energy in the European community: Supply and demand patterns for the medium term future. In Energy in the European communities (Frans A. M. Alting Van Geusau, ed.). A. W. Sijthoff, Leyden, Netherlands, 1975.

7. United Nations, Economic Commission for Europe. Long-term prospects of the electric power industry in Europe. New York, 1974.

8. International Research Group. Energy policies in the European community. Energy Research and Development Administration, Washington, D.C., 1976. See also Institute of Electrical and Electronics Engineers. The U.S. energy equation and the role of electricity. IEEE, Piscataway, N.J., 1976; G. Ray and C. Robinson, The European energy market in 1980. Intl. Pub. Serv., London, 1975; Organization for Economic Co-Operation and Development. Energy production and the environment. OECD, Paris, 1977; Organization for Economic Co-Operation and Development. World energy outlook. OECD, Paris, 1977.

9. M. P. Meinel and A. B. Meinel. National energy/gross national product trajectories. Unpublished paper presented at the International

Scientific Forum on an Acceptable Nuclear Energy Future of the World. University of Miami, Coral Gables, Fla., 1977.

10. U.S. Federal Energy Administration, Office of Conservation and Environment. Energy conservation study: Report to Congress. U.S. Government Printing Office, Washington, D.C., 1974.

11. R. A. Lowe. Energy conservation through improved solid waste management. Report SW-125. Environmental Protection Agency, Washington, D. C., 1974.

12. W. E. Franklin, D. Bendersky, W. D. Park and R. G. Hunt. Potential energy conservation from recycling metals in urban solid wastes. In Energy conservation papers (R. H. Williams, ed.). Ballinger, Cambridge, Mass., 1975, pp. 171-218.

13. A. Poole. The Potential for Energy Recovery from Organic Wastes. In ibid., pp. 219-308.

14. H. Alter. Energy conservation and fuel production by processing solid wastes. Environmental Conservation 4: 11-19 (1977).

15. H. C. Bailly and C. Tayart de Broms. Material flows in the post consumer waste stream of the EEC. Graham & Trotman, London, 1977.

16. Europool. Secondary materials in domestic refuse as energy sources. Graham & Trotman, London, 1977.

Chapter 3

QUANTITY AND COMPOSITION OF WASTES
IN THE UNITED STATES AND EUROPE

The quantities and composition of municipal solid waste differ in the
United States and Europe. The amount of available waste and its
makeup determine, in part, the type of recovery technology to be
adopted. These two factors are related to the affluence of the economy, socioeconomic patterns within the community, type of food distribution system, the season, and even the day of the week. Few of
these points have been definitively quantified, but some trends will be
presented in the following discussion. The quantity and composition
of municipal solid waste, more than other factors, determine the
contrasting approaches to resource recovery in the United States and
Europe. It is important to be aware of these characteristics to understand differences in recovery technologies.

WASTE GENERATION AND COMPOSITION
IN THE UNITED STATES

The composition and quantity of municipal solid waste generated in the
United States has been reported by the U.S. Environmental Protection
Agency [1]. Their estimates are summarized in Table 3.1. Note that
these are average estimates and do not necessarily represent the
waste from any particular community.

The agency has also estimated the per caput waste generation rate
in the past and projected its change over the next several years. These
estimates and projections are given in Table 3.2. Presumably, the
estimates were carefully made using the best available data. However,
they are not necessarily good future forecasts. For example, if in
1990 the generation is 5 pounds (2.3 kg) per person per day, and noting
that approximately half of this total is household waste, this means a
family of four would bring into their home more than 10 pounds of

23

TABLE 3.1. Material Flow Estimates of Residential and Commercial Postconsumer Net Solid Waste Disposed of, by Material and Product Categories, 1975

Material category	Product category (Millions of U.S. tons, as-generated wet weight)[a]							Totals			
	Newspapers, books, magazines	Containers, packaging	Major household appliances	Furniture, furnishings	Clothing, footwear	Food products	Other products	As-generated wet weight[a]		As-disposed wet weight[b]	
								Million U.S. tons	Percentage	Million U.S. tons	Percentage
Paper	9.8	19.1	tr.	tr.	—	—	8.3	37.2	29.0	44.9	34.9
Glass	—	12.2	0.1	tr.	—	—	1.0	13.3	10.4	13.5	10.5
Metals	—	5.9	2.1	0.1	—	—	4.0	12.1	9.6	12.6	9.8
Ferrous	—	(5.2)	(1.8)	(0.1)	—	—	(3.7)	(10.8)	(8.6)		
Aluminum	—	(0.7)	(0.1)	tr.	—	—	(0.1)	(0.9)	(0.7)		
Other nonferrous	—	—	(0.2)	tr.	—	—	(0.2)	(0.4)	(0.3)		
Plastics	—	2.7	0.1	0.1	tr.	—	1.5	4.4	3.4	4.9	3.8
Rubber and leather	—	tr.	tr.	tr.	0.7	—	2.6	3.3	2.6	3.4	2.6
Textiles	—	0.1	tr.	0.6	0.5	—	0.9	2.1	1.6	2.2	1.7
Wood	—	1.8	tr.	2.6	—	—	0.5	4.9	3.8	4.9	3.8
Total nonfood product waste:	9.8	41.7	2.3	3.4	1.2	—	18.9	77.5	60.5	86.5	67.3
Food waste	—	—	—	—	—	22.8	—	22.8	17.8	19.1	14.9
Total product waste:	9.8	41.7	2.3	3.4	1.2	22.8	18.9	100.3	78.3	105.6	82.2
Yard waste								26.0	20.2	20.9	16.3
Misc. inorganic materials								1.9	1.5	2.0	1.6
Grand totals:								128.2	100.0	128.5	100.0

[a]"As-generated" weight basis refers to an assumed normal moisture content of material in its final use prior to discard, for example: paper at an "air-dry" 7 percent moisture; glass and metals at zero percent. Total waste, including food and yard categories, estimated at 26 percent moisture.

[b]"As-disposed" basis assumes moisture transfer among materials in collection and storage, but no net addition or loss of moisture for the aggregate of materials. Based on estimates in W. R. Niessen and S. H. Chansky. The nature of refuse. In Proceedings, 1970 National Incinerator Conference. Cincinnati, May 17–20, 1970. ASME, New York, pp. 7–8.

Note: Net solid waste disposal defined as net residual material after accounting for recycled materials diverted from waste stream. Details may not agree with totals due to rounding.

Source: Ref. 1.

TABLE 3.2. Baseline Estimates and Projections
of Postconsumer Solid Waste Generation

Year	Total discards per person per day[a]	
	lb	kg
1971	3.52	1.60
1973	3.75	1.70
1974	3.70	1.68
1975	3.40	1.54
1980	4.28	1.94
1985	4.67	2.12
1990	5.00	2.27

[a]Based on annual waste generation for the country
and computed on the basis of 365 days per year.
Source: Ref. 1.

material each day. The magnitude of the forecast seems to challenge
common sense. Thus, projections of waste generation for planning
purposes should be used cautiously.

   The estimates in Table 3.2 of past waste generation are almost
constant—approximately 3.6 pounds (1.6 kg) per person per day. The
figures for past years show no trend of increasing waste generation,
which is another reason for using future projections with caution.
Indeed, there are some indications that per caput generation is de-
creasing, as manufacturers attempt to reduce costs by adopting packaging innovations. There may be merit in using retrospective figures
for prospective projections in the absence of better data. *

   The estimates of Tables 3.1 and 3.2 are arrived at by input-output
techniques [2]. Generally, the input data are information of the value
of production, translated into products. The output, or waste to be
disposed of, is arrived at by subtracting such categories as exports

---

*Retrospective data for the United Kingdom show little change in the
mass of waste generated, 0.762 kg/person/day in 1931 to 0.857 kg/
person/day in 1976. However, the density of the waste decreased
dramatically over this same time period, from 342 kg/m^3 to 150 kg/
m^3 [14].

and durables, allowing for the estimated life of the latter. The
limitations of such models have been addressed [2]. It is important
to note, however, that three attempts to estimate waste quantities
and composition in the United States are generally in agreement (see
Fig. 3.1). The results suggest that the various estimates may accu-
rately portray the average situation for the country. However, they
do not present waste generation and composition in specific communities.

The average quantities—or average per person generation rates—
may be grossly inaccurate in planning a waste-to-energy recovery
system in a specific community. These points are expanded in Chapter 7.

WASTE GENERATION AND COMPOSITION
IN EUROPE

The amount of materials disposed of in solid waste facilities has been
estimated for the European Economic Community (EEC) for paper and
board (dry basis), iron and steel (not including automobiles and bulky
items), aluminum (from households), glass, textiles, and rubber
tires [3]. Based on these estimates, the EEC population of 253 million persons generated 31 million metric tons of the materials listed.
This quantity is not directly comparable to the 128 million U.S. tons
(116 million metric tons) presented in Table 3.1, but nonetheless it
indicates that less material is discarded in Europe than in the United
States.

The energy equivalent of the discards in the EEC (more than just
those listed in Table 2.3) have been estimated to be 5×10^{17} joules [3].
A comparable estimate for the United States is 13×10^{17} joules [4].
A recent EEC survey [5] suggests that the amount of waste collected
is a function of the size of the community. Table 3.3 surveys the
amount of waste collected in West German communities. Similar observations have been made in the Republic of Ireland [5]. The differences in waste generation in various EEC countries are illustrated in
Table 3.4, which lists the quantity of waste per inhabitant per year for
various major cities. The values differ by more than a factor of 2.

Perhaps the differences illustrated by Tables 3.3 and 3.4 are due
to contrasting life styles and community sizes. In addition, people in
smaller cities may also dispose of some waste by means other than
municipal collection.

The reported composition of municipal solid waste differs among
countries and cities within Europe, especially the relative amounts of
metals, glass, paper and food wastes, as shown in Table 3.5.

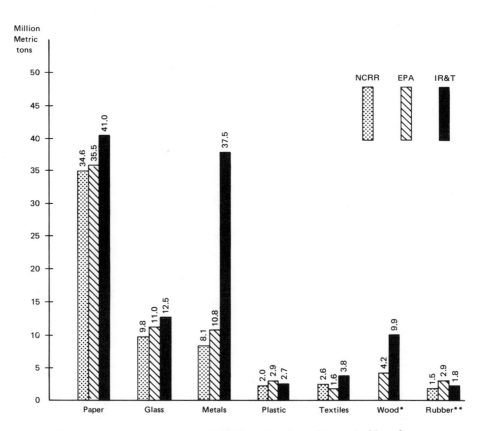

FIGURE 3.1. Comparisons of IR&T estimates of household and com-
mercial waste (1971) with NCRR and EPA studies. *No estimate for
NCRR. **EPA estimate combines rubber and leather. (Sources:
Municipal solid waste. NCRR Bulletin 3(2): (1973) Fred L. Smith.
Estimates of household and commercial solid wastes based on pro-
duction statistics. Draft report. Environmental Protection Agency,
Office of Solid Waste Management Programs, Washington, D.C.)

TABLE 3.3. Size of Community and Incidence
of Waste in Germany

No. of inhabitants in the community	Amount of waste per person per year	
	lb	kg
Below 2000	200	100
2000–5000	353	160
5000–20,000	397	180
20,000–100,000	441	200
100,000–500,000	485	220
500,000–1,000,000	529	240
Over 1,000,000	573	260

Source: Ref. 5.

TABLE 3.4. Quantity of Waste per Inhabitant per Year in Various
European Cities and Countries

Country/city	Quantity	
	lb	kg
Belgium	681	309
Denmark	675	306
France	518	235
Bordeaux	694	315
Germany	661	300
Great Britain (average values)	617	280
Edinburgh	463	210
Suburban areas with numerous gardens	551	250
Ireland	441	200
Italy	575	261
Luxembourg (city)	882	400
The Netherlands (average values)	595	270
The Hague	606	275
Groningen	551	250
Switzerland (for incineration only)		
Basel	362	164
Lausanne	419	190
Zurich	375	170

Source: Refs. 5 and 6.

TABLE 3.5. Comparison of Household Solid Wastes (Wt. %, as Received)

Component	Location									
	Holland [7]	Sweden [8]	Stockholm [9]	Stevenage (U.K.) [4]	Rome [11]	Hamburg [12]	Vienna [11]	Prague [11]	Sofia [11]	Madrid [13]
Paper	22.6	40-50	45	33	18	34.8	35.3	13.4	10.0	15-30
Iron and steel	3.1	4-8	6	7	3	4.2	9.7	6.2	1.7	2.5-6.0
Mixed nonferrous metals	0.1	-	0.5	<1	-		-			
Glass	13.0	8-10	7	10	4	15	9.1	6.6	1.6	2.5-10
Plastics	5.3	4-8	9	4	4	4.5	5.5	4.2	1.7	3-12
Textiles and rubber	2.6	2-6	-	4	-	6.1	5.6	8.1	7.0	1.3-12
Food wastes (garbage)	49.6	8-20	17	14	50	16.7	24.1	41.8	54.0	30-60
Other organic materials			8.5	-		-	-	-	-	-
Sand, stones, other inorganic materials	1.4	-	7	28	-	18.7	9.9	19.7	24.0	2.2-11

Note 1: In some instances, the columns do not total 100 percent. Not every component of the waste was always classified.
Note 2: Numbers in column headings refer to references at the end of this chapter.

Differences in composition may be related to standards of living and life styles, particularly with respect to the extent of centralized food processing and distribution and the use of household refrigeration. Surveys of waste composition in various parts of Europe which still burn solid fuels (e.g., coal or coke) from home heating may be misleading. As these fuels are replaced with oil or gas, the volume and character of household wastes change. In the future, more paper (and other combustible material) will likely be collected, which will have an effect on decisions to adopt various types of materials and/or energy recovery systems. Table 3.6 summarizes the methods of waste disposal used in some European countries [6]. (The numbers in this table differ, but only slightly, from some other reports, but such differences are to be expected considering the difficulty in accumulating such statistics.)

TABLE 3.6. Methods of Waste Disposal

Member country	Disposal as % of total waste stream		
	Landfill[a]	Incineration	Composting
Belgium (Flanders only)	62	29	9
Germany	72	25	3
Netherlands	50	30	20
U.K. (England only)	91	9	<1

[a]Includes shredding followed by landfill.
Source: Ref. 6.

NOTES AND REFERENCES

1. Resource recovery and waste reduction: Fourth report to Congress. Report SW-600. Environmental Protection Agency, Washington, D.C., 1977.
2. D. Kidder. Forecasting the composition and weight of household solid wastes using input-output techniques. Report EPA-600/8-77-002. Environmental Protection Agency, Cincinnati, 1977.

3. H-C. Bailly and C. Tayart de Borms. Material flows in the post consumer waste stream of the EEC. Graham & Trotman, London, 1977.
4. H. Alter. Energy conservation and fuel production by processing solid wastes. Environmental Conservation 4: 11-19 (1977).
5. Europool. Secondary materials in domestic refuse as energy sources. Graham & Trotman, London, 1977.
6. M. Webb and L. Whalley. Household waste sorting systems. Commission of the European Communities, Brussels, 1977.
7. B. G. Kreiter. Composition of waste and some possibilities of recovery. Conservation and Recycling 1: 19-30 (1976).
8. Households contributing to resource recovery. Swedish Institute for Resource Recovery, Malmo, 1975.
9. B. Citron and B. Halén. Automated recovery from domestic refuse. Report STU 73-5182. AB Svenska Fläktfabriken, Stockholm, 1978.
10. E. Douglas and P. Birch. Recovery of potentially re-usable materials from domestic refuse by physical sorting. Resource Recovery and Conservation 1: 310-44 (1976).
11. Sorain Cecchini Sp. a. Via Bruxelles 53, 00198 Rome (untitled and undated).
12. H. W. Kindler. Müllsortieranlagen: Wistschaftlichkeit und Standortprobleme. Chemie Analgen u. Verfahren 1 (1976).
13. M. M. Cavanna, E. Riaño, J. Sanchez Almarez, and H. Garcia Ramirez. Latest developments in processing Spanish urban raw refuse. In Proceedings, Fifth Mineral Waste Utilization Symposium (E. Aleshin, ed.). IIT Res. Inst. and Bureau of Mines, Chicago, 1976, pp. 141-45.
14. F. Flintoff. Quoted in Solid Wastes 68(9): 354 (1978).

Chapter 4

RECOVERY OF MATERIALS FROM
WASTE IN EUROPE

The technology of materials recovery from solid waste in the United States has been described in several publications and is not repeated here. However, there are few descriptions of parallel developments in Europe beyond incineration and composting. European countries share with the United States such concerns as increasing amounts of solid waste, diminished land areas for disposal, and public interest in materials recovery and energy conservation.

The technological approaches to materials recovery in Europe are different from the United States in several important aspects. First, in Europe there is an emphasis on paper recovery, since many countries anticipate future fiber shortages. Second, European waste generally contains smaller quantities of metal than in the United States. Similarly, there is less interest in mechanized glass recovery in Europe. Consequently, gross revenues from selling recovered materials would be low, except for the potential of paper recovery. Third, there is less waste generated in European cities than in American communities of similar populations. These reasons dictate different kinds of recovery technology. For example, European practices generally do not rely on massive hammermills as the first step in waste processing. The amount of waste available simply cannot support the capital cost of such devices. As a result, small communities in the United States can learn from European experiences.

The following discussion expands and updates previous reviews [1-3]. Because the emphasis on paper recovery is so different from practices in the United States, a subsequent section summarizes some published reports of the properties and use of paper furnishes recovered from mixed municipal wastes.

32

ENGLAND

The Warren Spring Laboratory (United Kingdom Department of Industry), with support from the Department of the Environment, operated a research pilot plant for the physical sorting of household refuse to recover paper, magnetic metals, glass, and refuse-derived fuel (RDF) [4]. A small (< 1 percent) amount of nonferrous metals, as well as mixed plastic film and sheeting, could also be recovered. The pilot plant operated at 3 to 4 tons per hour. Several processing circuits have been investigated; the one shown in Fig. 5.1 is a typical example. The materials balance has been reported [4].

In this process, household refuse is conveyed to a device that opens plastic and paper bags to liberate the refuse. (It is adapted from a type of screw compactor used on collection trucks.) The refuse is sized through a rotary screen (trommel) which separates it into four fractions. Fine material (< 13 mm) is discarded, middling material (50 x 13 mm) is magnetically separated, and putrescible material in the nonmagnetic portion is ballistically separated from denser glass and stones by bouncing the mixture off a rapidly rotating drum and by sink/float in brine. Glass can be recovered from the sink fraction.

The other middling fraction (200 x 50 mm) is similarly processed, except the organic materials from the ballistic thrower are air-classified to produce an RDF. The air classifier used at Warren Spring was a rotating drum much like some used in the United States [5]. However, there is a screen section in the drum to remove the fines. (A scale-up of this drum to 10 tons per hour is 3 m in diameter and 9 m long.) The light fraction is the RDF.

The trommel oversize (> 200 mm) is shredded and air-classified for paper recovery. This yield is about 20 percent of the input waste and would contain 6 to 8 percent contraries, mostly plastics. The sand, glass, and food waste in the recovered paper is reduced by the trommel screening; the plastic film and textiles can be reduced by two other processes.

As shown in Fig. 4.1, "flexibles" are removed in the trommel. This is done by a constantly moving trolley wire through the trommel near its top. Rags and plastic film drop over the wire which carries these materials away. This separates rags and plastic film from the final paper product and reduces "blinding" of the trommel holes. Plastic film is also removed from the trommel oversize product with a blower that is actuated by a laser beam. The specular reflectance

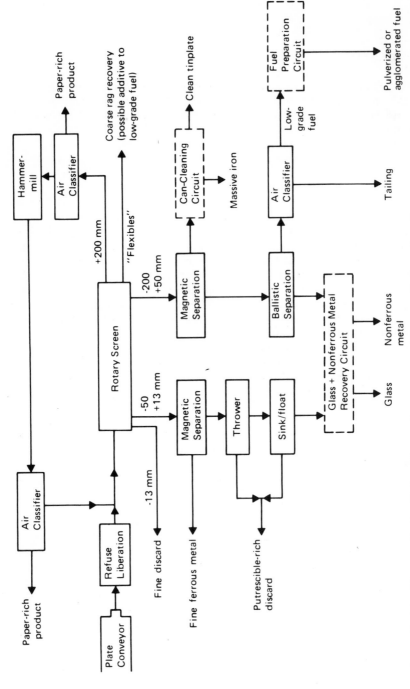

FIGURE 4.1. Warren Spring Laboratory primary raw refuse separation circuit.

34

from the laser distinguishes between plastic and paper; the wavelength
of the laser is crucial to proper separation.

The development work at this pilot plant has led to two commercial (municipal) plants, one in Doncaster (South Yorkshire) and another
in Byker (Tyne and Wear, Northumberland). These are joint ventures
of the local authorities, the Department of the Environment, and the
Warren Spring Laboratory.

The Doncaster plant has an initial design capacity of 10 tons per
hour per line. Initially, one line is installed, and it is planned to add
a second later. With two lines, it is conservatively estimated the
plant will process 1250 tons per week, operating two shifts per day.
It is believed that as selected commercial wastes are received, the
daily throughput will be 320 tons. The plant is anticipated to recover
17,500 tons of material from an annual throughput of 62,500 tons. Recovery will include 5 percent magnetic metals, 16 percent RDF, 2 percent paper as secondary fiber, and 5 percent glass product [6]. The
last will not be recovered for reuse in the manufacture of containers,
at least not initially. The full process flow is shown in Fig. 4.2.

The Byker plant has as its objectives the recovery of densified
refuse-derived fuel (d-RDF) and magnetic metals. It is designed to
process 1500 tons per week of refuse. The process flow, Fig. 4.3,
resembles United States practices in that the waste is first fed to a
primary shredder. The unique feature is the use of a trommel before
the air classifier to remove ash components (<12 mm) and oversize
pieces (>150 mm) such as textiles which could interfere with the operation of the densifier. Similarly, the magnetic separator removes
small amounts of light-gauge magnetic metal from the light fraction
which could compromise the densifier's operation. The plant has been
designed to process domestic and commercial waste at a maximum rate
of 30 tons per hour. A similar process flow and preparation of d-RDF
on a pilot scale in the United States has been described [7].

The air classifier for the Byker plant was developed by Newell
Dunford Engineering, Ltd. It is basically a rotating cone with its
axis at a low angle to the horizontal. The feed enters the cone at its
wide end and falls into an induced flow of high-velocity air introduced
from around the periphery of the narrow or exit end. The light materials are deentrained in a plenum chamber. The heavy materials, such
as metals and large pieces of glass and rock, are cleaned as they tum-
ble in the drum in the turbulent airstream and leave the cone at its
narrow end. The prototype device operated at up to about 7 tons per
hour; the cone was 2.3 x 4.7 m in size.

The d-RDF from Byker will be burned in a 4.5-MW-capacity
boiler to supply district heating, ultimately for 2500 dwellings. The
first boiler unit was scheduled to be on-line in October, 1979. The
processing plant was undergoing commissioning trials in June 1979.

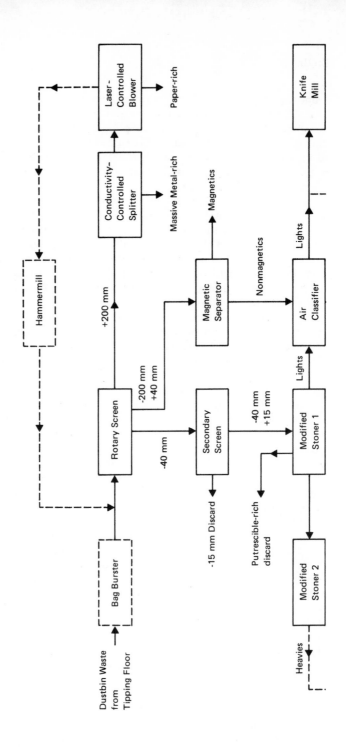

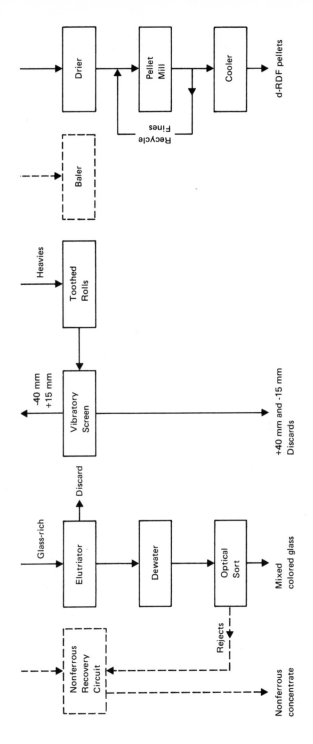

FIGURE 4.2. Doncaster refuse-processing plant: outline flowsheet.

37

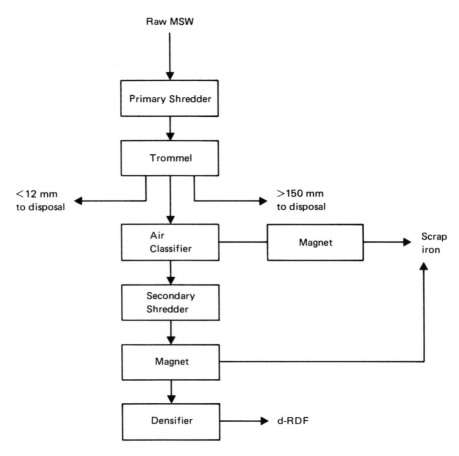

FIGURE 4.3. Flowsheet of the Tyne and Wear (Byker) plant.

    Also in England, the Department of the Environment and Newell
Dunford Engineering, Ltd. are cooperating in building an experimental
plant of approximately 100-ton-per-day capacity for the recovery of
d-RDF and pulped paper fiber. The plant is located in Chichester,
Sussex. The process flow is shredding followed by rotary screening.
The fine materials (< 12 mm) and oversize (> 150 mm) are discarded.
The middling fraction (150 × 12 mm) is air-classified in the same
rotary cone device developed for Byker. A portion of the light fraction is reshredded and converted to d-RDF. Alternatively, the light

fraction may pass through a rotary screen (to remove material < 25 mm) and a second rotary drum air classifier. The latter's purpose is to separate plastic films and other contraries by wetting the feed to the air classifier and recovering the paper as a wet heavy fraction. This wet material is then fed to a series of processing steps used in the pulp and paper industry to produce a marketable refined pulp. This plant should be operating in 1979.

A private firm in England, Refuse Derived Fuels, Ltd., has been experimenting with an RDF preparation method in Burgess Hill, Sussex [8]. Shredded waste is magnetically scalped and then screened through a series of disc screens. The fines are discarded; the oversize is a fuel; and the moist middling fraction contains food waste. The middling is dried by partially composting it in a drum for about 24 hours. The heat from the biological reactions drives off some of the water and the remaining product is combined with the disc screen oversize or fuel fraction. In this way, there is a higher yield of drier fuel than would have been achieved otherwise. Furthermore, the drying is achieved without use of fossil fuels.

GERMANY

There have been several noteworthy resource recovery efforts in Germany. For example, a private contractor investigated shredding, air classification, and magnetic metals recovery for the Berlin street-cleaning department (BSR). Household and office wastepaper (from separate collection) has been used as 100 percent of the furnish to manufacture a kraftlike paper in Berlin. A plant was built in Berlin to make aggregate for concrete (called "sinter pumice") from inciner-ator residue, but the product's performance was less than expected and it operates intermittently, if at all. Resource recovery research is underway at the Technical University of Berlin. The objective of these activities is to decrease the need for disposal sites in land-short West Berlin.

A materials separation process was developed through the pilot stage at the Institute for Processing of the Rhine Westphalia Technical University in Aachen [9]. The pilot plant operated at from 1 to 1.5 tons per hour. The research phase was concluded in 1977, and the firm of Siebtechnik GmbH of Mühlheim/Ruhr was commissioned to bring the plant to an operational scale.

The process flow consists of sieving the refuse at 20 mm, mag-netic separation, sieving at 40 mm to prepare a fraction for compost, followed by a crusher or knife-type cutter rather than an impact-type mill. The crushed material is air-classified in a novel device that increases the time the material spends in the classification zone. The light fraction is collected as a mixture of paper and plastics and the heavy fraction processed further. The heavy fraction is divided into more and less dense fractions by a countercurrent water separator. The dense fraction again is magnetically separated followed by heavy media separation which is intended to remove ceramics from glass, so that it can then be optically sorted [10]. This process resembles other processes described in the United States [11], and judging from reports of both countercurrent water and heavy media separations, it appears unlikely that enough of the ceramics could be removed from the glass so as to meet specifications for reuse in container manu-facture [12, 13]. Also consistent with other published reports [14], the group at Aachen recognized that care must be exercised in the choice of the type of crusher or other mill used, otherwise the yield of glass pieces large enough to recover and color-sort is too low [10].

Research at the University in Hamburg has been directed at py-rolysis of waste plastics. They report using both fluidized bed and molten salt reactors [15].

In Munich, Babcock Krauss-Maffei-Industrielagen operates a 5-ton-per-hour pilot plant, called System R-80, for paper and metal recovery [16]. One fraction is suitable for compost and potential glass recovery. The process flow is shown in Fig. 4.4. There is no primary shredding, as with several other European plants.

As shown in Fig. 4.4, magnetic metals are recovered from the raw waste. Such scrap must be washed (or otherwise cleaned) prior to use. The nonmagnetic fraction is screened to remove compostable organic material (principally food waste), glass, and sand. The over-size is air-classified to a light fraction of paper and plastics, which is then screened to remove a mixture of plastic film and large paper sheets as the oversize product. The undersize product is air-classified to remove inorganic materials as a heavy fraction (which goes to waste) and a light fraction consisting of newspapers, magazine stock, and tex-tiles. As shown in Fig. 4.4, the paper fractions can be combined. The recovered paper contains 5 percent plastics and other contraries. Fig. 4.4 also shows the materials balance. Additional details have been published [10].

Babcock Krauss-Maffei has experimented with separating the plastic film from the paper and using it, mixed with fillers and/or industrial plastic scrap, to make extrusion products. Such extrusions are weak without additional cleanup of the plastics. The use of the R-80 recovered paper as a furnish to a paper mill has been investi-gated, and the results are discussed later in this chapter. Babcock

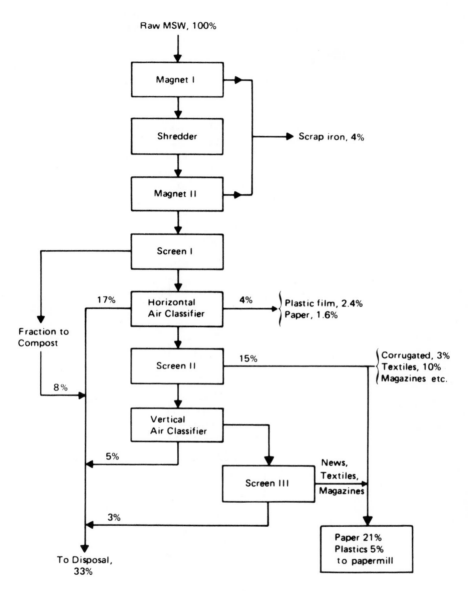

FIGURE 4.4. Process flow and materials balance of the Babcock Krauss-Maffei system.

Krauss-Maffei built an R-80 plant for the municipality of Landskrona, in the southern part of Sweden. Its capacity is 200 tons per day; a primary product is compost.

HOLLAND

In Haarlem, a private firm (Esmil-Habets) and a government laboratory (Central Technical Institute, TNO) are jointly investigating the separation of municipal solid waste and paper recovery [17]. The reported process flow for the 15-ton-per-hour plant is shown in Fig. 4.5. A distinguishing feature of this system is the use of two zigzag

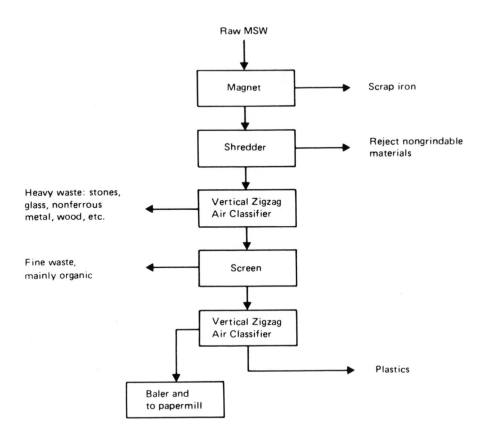

FIGURE 4.5. Process flow of the TNO system.

air classifiers so that the light fraction from the first is the feed for
the second (with screening in between). The objective is to separate
the plastics from the paper in the light fraction from the first classifier to produce a usable paper furnish, which would be the heavy fraction in the second classifier. To achieve separation, the moisture
content of the mixture of paper and plastics is raised so that the paper
becomes the heavy fraction. Reportedly, separation of Haarlem waste
resulted in 3 percent iron, 25 to 35 percent heavy fraction, 20 to 30
percent paper, and 5 percent plastics [18]. These results represent
a high recovery of paper if the Haarlem waste has a composition
similar to other parts of Holland [19]. The various products recovered in this pilot plant have been described elsewhere [20].

In addition, the Dutch government-owned waste management firm
VAM is constructing a 200-ton-per-day steel and paper recovery plant
at its composting and disposal site near Wijster. The plant is supplied
by the Swedish firm Fläkt and was scheduled for commissioning in
early 1980.

SWEDEN

In Sweden, a division of PLM, a packaging firm, has experimented
with shredding, magnetic separation, and the preparation of d-RDF
using a process developed in the United States (Papakube). A pilot
plant for the recovery of usable paper furnish was operated in Högdalen
in south Stockholm by A. B. Svenska Flätfabriken with support from
the Swedish Board of Technical Development. The Fläkt process is
shown in Fig. 4.6.

Household waste is shredded in a flail mill. The material is then
sized through a trommel. Material >200 mm is reshredded and the remainder fed to a zigzag air classifier. Magnetic metals are recovered
from the heavy fraction. Presumably, the heavy fraction could be processed further for glass and/or aluminum recovery.

The light fraction is reshredded in an impact machine with light-
weight, hinged beaters on two rotors, similar to the flail mill used for
the primary shredding. The light fraction is sized through a second
trommel. The undersize portion (<25 mm) contains most of the putrescible materials and is discarded. The oversize fraction (>80 mm) is
rich in plastic film. The middling (80 x 25 mm) is processed through
a pulp flash dryer (a commercial item manufactured by Fläkt) at 130°
C air temperature. This dries the paper and shrinks the remaining
plastic film, which is later separated by a cyclone. Textiles fall in
the pulp dryer and are removed. Some of the heat-shrunk plastic film
is not removed by the cyclone but can be removed in a subsequent air

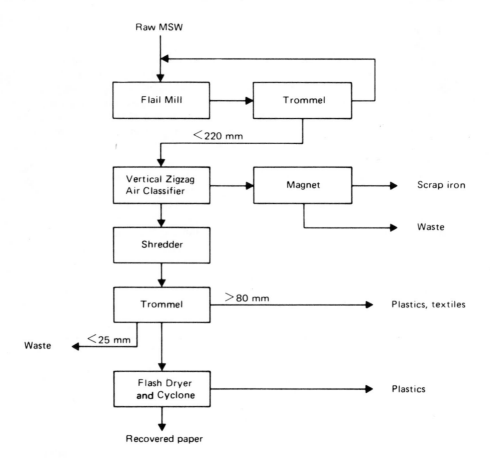

FIGURE 4.6. Process flow of the Fläkt system.

classifier which is reportedly capable of separating newspaper from cardboards [21].

Based on the investigations at the pilot plant, Fläkt has contracted to build a full-scale, 200-ton-per-day processing plant for the City of Stockholm. As mentioned earlier, Fläkt is also building a similar full-scale plant in Holland.

FRANCE

In association with an industrial group, the French Bureau de Recher-
ches Géologiques et Minières (BRGM) is researching materials recovery from municipal solid waste [22]. A pilot plant with a capacity
of $1\frac{1}{2}$ to 2 tons per hour is operated in Orleans. The process flow consists of first screening the refuse to separate three fractions: flexibles
(textiles and plastic films); flat cardboard, newspapers, and magazines;
and large packages. The flexibles are removed by a specially designed
conveyor, similar to the Warren Spring trolley wire, running along the
top of the trommel screen. The other separations are achieved by
selection of the trommel hole size and shape. The fraction larger than
200 mm is processed to remove magnetic metals, screened to eliminate < 40-mm fines, and air-classified to separate paper and plastic
film. Plastics are removed from the paper fraction by electrostatic
separation. Plastic bottles (principally polyvinyl chloride, PVC, in
France) are separated on an air table. Plastic scrap, glass, and
metals in the middling fraction are withdrawn from paper products
(principally hygienic papers) by differential rebound and adhesion.
This is achieved by dropping materials from one conveyor to another
which moves in an opposite direction. The second conveyor is kept
wet and the paper, foodstuffs, and so on adhere and are carried off
[22].

The first reports [22, 23] give the materials balance and describe
the BRGM process for separating incinerator residues and mixtures of
glass and plastic bottles. Projected plant costs (in France) are also
given. There are plans to build a full-scale plant (200 tons per day)
based on the BRGM process, in 1980 in Nancy.

In Tournan-en-Brie, the firm SOCEA operates a paper recovery
plant designed in cooperation with Empressa National Adara de Investi-
gaciones Mineras, Madrid. This Spanish group has been working for
several years in cooperation with the United States Bureau of Mines.
The SOCEA plant shreds the waste in a flail mill, air-classifies,
screens at between 20 and 50 mm, adds water, and feeds the material
into a "lacerating trommel" (which may resemble a rasp mill) to produce a paper furnish. The wet pulp is forced through the trommel
holes and rags and plastics are rejected. Cardboard is separated from
the rags and plastics in a type of ballistic separator and the cardboard
is recycled to the trommel to be included with the wet paper.

The plant serves 30 townships which supply 95 tons per day of
solid waste. As of May 1978, about a year after the plant was commissioned, the paper product had not yet been sold, but had been successfully made into felt papers. Reportedly, one reason for the lack

of sales was that the product was at times contaminated with grass clippings and similar wastes.

In Laval, there is a densified refuse-derived fuel plant constructed by Société d'Études et d'Ingénierie (SOCETING). The process is known by the name which has been given to the pelletized d-RDF, COMBOR (Combustibles d'ordures ménagères). The plant was built on the site of a shredding and composting station and uses the existing vertical hammermill to reduce the waste to 50 percent smaller than 50 mm. The shredded material is magnetically separated and screened at 20 mm to remove mostly inorganic material (25 percent of the feed is rejected as < 20 mm). The oversize is conveyed under an aspirator hood, the heavy fraction (about 1 percent) recirculated to the hammermill, and the light fraction pneumatically conveyed for further processing. The light fraction is deentrained in a cyclone, past a magnetic drum, and through a trommel screen to remove material < 8 mm (glass and stones). At this point, a 1.5 percent calcium carbonate is added in a storage and mixing hopper, ostensibly to neutralize hydrogen chloride released during combustion (from the high PVC content of French waste). In the storage hoppers the material is intentionally permitted to ferment or compost as a means of driving off moisture. It is then pelletized in two small presses which are fitted with stationary cylindrical dies made up of two semicircular sections. The pellets are 16 mm in diameter and approximately 25 mm long. The capacity of the plant is reportedly 10,500 tons per year, and the fuel would supply 63 percent of the capacity of a local district heating system when the system is operated 8 hours per day. Other details have been published [24], including a report of the fuel properties. However, the reported properties are inconsistent (low ash, but also low ash fusion temperature, as an example) and must be discounted.

ITALY

Recovery technology in Italy has to be responsive to large fluctuations in composition with season of the year, regions, and size of community, as well as large differences in per caput generation across the country. The national average per caput generation is 0.72 kg per day, varying from 0.52 to 0.82 kg [25] in different regions. The variations in composition have been reported; for example, the content of putrescible materials varies from 15 to 20 percent in large cities and from 15 to 35 percent across the country [25].

In Italy, 33 percent of the waste is incinerated, 9 percent is treated by plants employing both composting and incineration (no distinction is reported whether incineration is with or without heat recovery), 1 percent is composted, and 5 percent is processed through four recovery plants. The latter are in:

Perugia (built by the Cecchini Co.)	100 tons per day
Rome (built by the Sorain Co.)	650 "
Rome (built by the Cecchini Co.)	650 "
Rome (built by the SARR Co.)	550 "

All three of the Rome plants now belong to the Sorain-Cecchini Company [26]. The fourth is owned by the City of Perugia. A generalized process flow for the Sorain-Cecchini process is shown in Fig. 4.7.

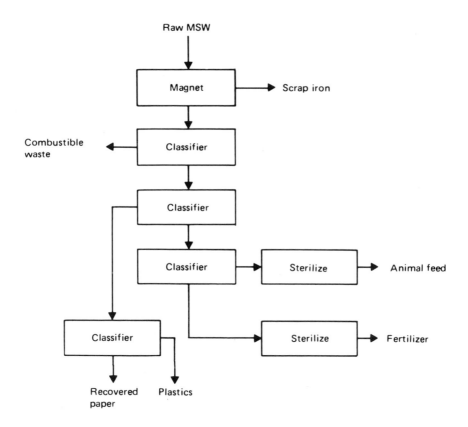

FIGURE 4.7. Process flow of the Sorain-Cecchini system.

The Sorain-Cecchini process can be illustrated by details of the plant in Perugia. Although initially small, it was reported in 1977 that the plant was being enlarged to 220 tons per day [25]. Other details are from this reference.

The composition of the waste in Perugia and the output products of the plant are given in Table 4.1. To achieve this, the "sorting unit" works two shifts and the "fodder, scrap iron, and incineration" units work three shifts. A detailed process flow diagram is shown in Fig. 4.8. After the refuse bags are opened, the material is conveyed onto a vibrating sieve with bars and steps designed for this purpose. The fine organic material separated here is sent directly to compost. The keys to the operation are the various proprietary classifiers, which include trommel screens and a "ragger."

TABLE 4.1. Percentage Composition of Input Waste and Outputs of the Processing Plant, Perugia, Italy

Composition

Paper	22%
Ferrous scrap	3–3.5
Plastic	4
Glass	2–2.5
Gross organic matter	27
Fine organic matter	15
Unclassifiable inert	26–27

Output products

Paper	12%
Animal fodder	5 (moisture 5–7%)
Ferrous scrap	3–3.5
Compost	15 (or 20% with ash recovery)
Plastic	3.5
Incinerated matter	35
Moisture	30

Source: Ref. 25.

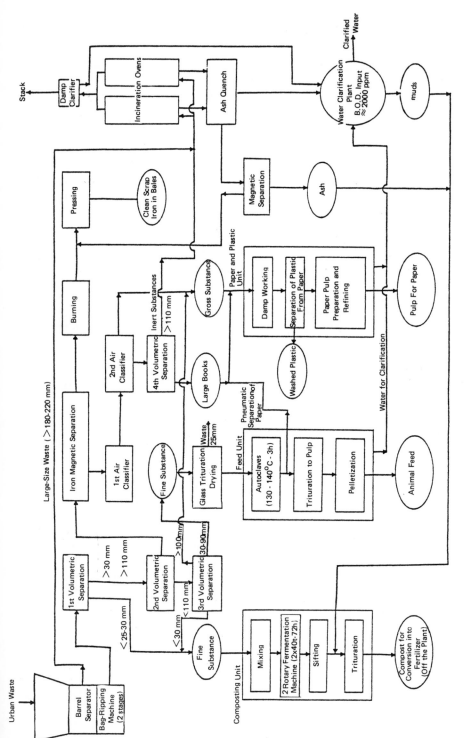

FIGURE 4.8. Process flow of the Sorain-Cecchini Perugia urban-waste-recycling plant. (Source: Ref. 25.)

The animal fodder is sterilized at a temperature above 140°C at a pressure of 4 atmospheres for more than 2 hours in a stirred autoclave. This step achieves some drying of the product. The steam for the autoclaves is obtained from the heat recovered from the flue gases from incineration of the nonreusable materials. The output of the autoclave contains some pieces of glass, wood, plastic, and so on, which are removed by screens, pneumatic separation, and density separation.

The recovered paper is pulped on-site, which gives an opportunity to remove foreign materials, such as plastic film. The pulp would be a good medium for bacterial growth and composting, so presumably is used promptly.

Few other details of the Perugia or Rome plants have been reported. However, in conjunction with the Rome plants, it is reported that Sorain-Cecchini was building a plant to reprocess the recovered plastic (to new refuse bags) in 1977. Also, the compost produced had nil commercial value. The animal feed analyzed 13 to 15 percent protein, 8 to 10 percent fats and lipids, and 10 percent ash, with a nutritional value of 60 percent of that of corn (maize) and was "supplied to cattle farms" [25].

It was also reported [25] that the paper pulp recovered in Rome was sold to paper mills at a price linked to that of corrugated paper and that the magnetic metals were cleaned by "baking" and sold baled at prevailing scrap prices. Interestingly, much of what has been written about the Sorain-Cecchini plants points out that the plants are "automatic," but it is also reported that one plant (Ponte Malnome, Rome) employs 1 manager, 10 office staff, and 67 workers; 58 people work shifts [25].

Sorain-Cecchini plants were reportedly under construction in Venezuela and Yugoslavia during 1976. Licenses for the process have been granted to firms for the United States, Canada, and Great Britain.

In Milan, the firm De Bartolomeis operates a 10- to 12-ton-per-hour pilot plant. The process flow begins with a precrusher to open bags, followed by magnetic separation. The waste is then passed through a rotary screen to remove putrescible materials and stones which are processed on an air table to separate the organic portion for composting. The oversize, representing approximately 70 percent of the waste, is processed in a horizontal air classifier to separate paper and plastic film. The remainder is shredded and passed through a second air classifier: the light fraction is a refuse-derived fuel.

Also in Italy, in 1977 Fiat was building its own 75-ton-per-day industrial waste-recycling plant, using technology of the sort pioneered by the U.S. Bureau of Mines. Siet/Tecneco, a company which builds incinerators (as does De Bartolomeis), also has a system for shredding and air classification for the recovery of paper and refuse-derived fuel.

USE OF RECOVERED PAPER

The use of paper fiber recovered from mixed municipal waste is a relatively new concept. The practice is not always enthusiastically received because it seemingly shifts the burden of waste disposal from a municipality to a papermaking plant. However, much of the research and development activity in European paper recovery has focused on removing contraries and dirt from the recovered paper so that it can meet specifications as a useful furnish for papermaking plants. Because much of this work has only recently been reported, it is worthwhile to summarize some of the results. Perhaps this will add to the credibility of the European materials recovery efforts and help introduce the concept of paper recovery from mixed wastes to the United States.

Almost all paper recovered from the various processes described in this chapter contains contraries, such as plastics and textiles. For example, the Warren Spring paper-rich fraction may contain 87 percent paper, 5 percent plastic, and 3 percent textiles [4]. It may include less depending on how well the removal devices are working—the trolley wire and laser detector.

Presumably, many of the contraries could be removed by processing the recovered material with pulp-cleaning equipment, either in a recovery facility (as in Chichester or Rome) or a papermaking plant. But such processing adds cost. Depending on composition, paper recovered from municipal waste may be usable, but not necessarily salable, without further processing.

In addition to contamination by textiles and plastics, the recovered paper may be mixed with food wastes. Also, it is likely to be wet, thus providing a medium for bacterial growth and composting. An exception is the paper recovered by the Fläkt process, which may be pasteurized as a result of drying. Cellulolytic bacteria and fungi are dormant when the paper contains less than about 9 percent water. Paper recovered by the Fläkt process exits the flash dryer containing about 2 percent water.

Extensive papermaking trials have been conducted with paper recovered by Babcock Krauss-Maffei. Some properties of the paper produced in these trials are summarized in Table 4.2 [27]. It was reported that the drying in the papermaking machine can be sufficient to kill residual bacteria [28, 29]. However, temperatures around 72 °C may cause a horny skin to form on the paper fibers [29]. The paper

TABLE 4.2. Comparison of Papers Manufactured from Recovered Fiber

Property	Mixed wastepaper	Household wastepaper	Furnish from: Black-Clawson, Franklin, OH	Cecchini, Rome	Babcock Krauss-Maffei, Munich	"Splinter-free" furnish from: Black-Clawson, Franklin, OH	Cecchini, Rome	Babcock Krauss-Maffei, Munich
Degree of pulping (°SR[a])	40	50	24	39	39	28	45	44
Splinters (%)	2	1	7	4	2	0	0	0
Long fibers (%)	38	34	39	29	36	45	15	37
Short fibers (%)	34	21	26	26	25	28	35	25
Fines (%)	26	44	28	40	37	27	50	38
Tear length (m)	2700	2400	1650	1300	2005	2300	1700	2400
Tear strength (g/cm)	9.8	8.8	8.5	7.3	9.0	8.5	5.4	10.0
Bend stiffness (N x mm)	0.39	0.41	0.33	0.25	0.35	0.34	0.25	0.36
Bursting strength (ko/cm^2)	1.20	—	—	—	1.25	—	—	—
Reflectance	40	42	30	35	38	30	34	38

[a]°SR = Schopper-Riegler, also called "freeness value."
Source: Ref. 27.

recovered at the Babcock Krauss-Maffei pilot plant was used in a paper mill to manufacture a grey board (density 600 g/m^2). The physical, chemical, and sensory tests showed no significant deviations between this board and ordinary production, except a small decrease in bending rigidity and tensile strength when the proportion of recovered paper was increased. The range of 10 to 50 percent recovered paper was investigated [29].

The paper fraction from the TNO pilot plant (Haarlem, Holland) was used to make a grey board (schrenz papier) in four so-called semitechnical and three industrial paper plants [20]. In the industrial plant, about 15 tons of paper were made from fibers recovered in the Haarlem plant. Three tons of the product were converted to corrugated cardboard (liner and fluting). Its properties are given in Table 4.3 and compared to similar paper made from ordinary mixed wastepaper. There is no evidence of harmful microbiological growth [20].

The Fläkt group has also supplied recovered paper for papermaking trials [21]. Table 4.4 shows some property values of papers made with Fläkt pulps compared to products made with more traditional sources of pulp. These and similar results indicate that furnishes A-1 and A-2 are probably suitable as additives in the intermediate layer of paperboard because its high specific stiffness is similar to materials made from virgin pulp. It is also much like paper made from the TNO pulp [20].

The Fläkt investigators report 86 to 97 percent reduction in the microorganism count of the recovered paper after heating the paper to 150°C. They do not indicate any observation of a horny skin forming on the fibers, as mentioned before.

The foregoing three investigations of using recovered paper all report that heat treatment at about 100°C will kill most microorganisms in the paper, which is not surprising. However, few reports relate this observation to normal papermaking practices or portray microorganism counts in normal production paper, or other control values. The bacteria count in the backwater of paper mills is reportedly between 10^5 and 10^7 per ml, but this has not been related to the paper produced [21]. The content of microorganisms in separately collected newspaper can vary widely, from almost none to what seem like high values, but these results have not been related to papermaking practices [30]. Consequently, the implications of microorganism counts are presently unclear.

There is concern for the extent of microbiological contamination of papers made from furnishes recovered from mixed municipal waste. Obviously, more work is needed. Based on the properties of papers made from such furnishes, they are likely to be used to make products

TABLE 4.3. Comparison of Paper Properties — TNO/Esmil

Property	Paper manufactured from paper fraction from the TNO/Esmil recycling separation system			Mixed waste paper	
	Three trials			Two trials	
Freeness ($\circ$ SR)[a]	27.5	45	74	55	50
Basis weight (g/m^2)	150	114	121	100	122
Sheet thickness (μm)	334	250	242	220	220
Air permeability (ml/min)	—	635	345	—	430
Bursting strength (kPa)	127	110	122	78	160
Tearing strength (N) md[b] md[b]	0.87	0.57	0.63	0.38	0.77
cd[b] cd[b]	0.88	0.64	0.81	0.45	0.87
Tensile strength (N/15 mm) md[b]	66	45	56	38	64
cd[b]	32	20	21	22	33
Stretch at break (%) md[b]	1.2	1.2	1.4	1.4	1.4
cd[b]	1.5	2.0	2.8	1.7	2.2

[a] $\circ$ SR = Schopper-Riegler, also called "freeness value."
[b] md = machine direction; cd = cross direction.
Source: Ref. 20.

such as grey board and corrugated. These finished papers will not
(and probably should not) come in direct contact with foodstuffs.

The paper recovered from municipal waste consists of a mixture
of mechanical and chemical fibers, clay and other fillers, as well as
impurities. The final mixture will largely determine the products to
be made. The interest of paper mills will probably depend on the
nature of their current products as well as the equipment available for
processing recycled paper.

TABLE 4.4. Comparison of Paper Properties — Fläkt

Property	Mechanically sorted		Manually sorted		Virgin
	Fläkt A-1 mixed[a]	Fläkt A-2 light[a]	Newsprint	Magazine	Mechanical pulp
Freeness ($°SR^b$)	37	46	62	71	51
Tensile strength (Nm/g)	22.7	21.1	36.2	36.6	19.1
Modulus of elasticity (GN/m^2)	1.74	1.53	2.06	2.84	1.33
Bending stiffness $[Nm/(kg/m^2)^3]$	1.12	1.24	0.96	0.71	1.22

[a]Furnish A-1 represents mixed paper, whereas A-2 is a fraction of lighter substances, predominantly mechanical fibers.

[b]$°SR$ = Schopper-Riegler, also called "freeness value."
Source: Ref. 21.

NOTES AND REFERENCES

1. H. Alter. Resource recovery in Europe. NCRR Bulletin, Fall 1976, pp. 77-84.
2. H. Alter. European materials recovery systems. Environ. Science and Technology 11: 44-48 (1977).
3. W. D. Conn. European developments in the recovery of energy and materials from municipal solid waste. Report EPA-600/7-77-040 (PB 270 219). Environmental Protection Agency, Cincinnati, 1976.
4. E. Douglas and P. Birch. Recovery of potentially re-usable materials from domestic refuse by physical sorting. Resource Recovery and Conservation 1: 319-44 (1976).
5. M. R. Grubbs, M. Paterson, and B. M. Faubus. Air classification of municipal refuse. In Proceedings, Fifth Mineral Waste Utilization Symposium (E. Aleshin, ed.). IIT Res. Inst. and Bureau of Mines, Chicago, 1975, pp. 169-74; M. R. Grubbs and E. J. Coulombe. Evaluating rotary drum air classification of shredded solid waste. In Proceedings, First Recycling World

Congress (M. Henstock, ed.). Exhibitions for Industry, Oxted, Surrey, 1978, Vol. 4.

6. The Doncaster project. Solid Wastes 67: 573-82 (1977).

7. H. Alter and J. M. Arnold. Preparation of densified refuse-derived fuel on a pilot scale. In Proceedings, Sixth Mineral Waste Utilization Symposium (E. Aleshin, ed.). IIT Res. Inst. and Bureau of Mines, Chicago, 1978, pp. 171-77.

8. Refuse Derived Fuels, Limited, London, N17 ODH, England.

9. H. Hoberg and E. Schulz. Process and apparatus for sorting refuse. U.S. Patent 4,043,513 (August 23, 1977).

10. Umweltbundesamt. Household waste sorting systems in the Federal Republic of Germany. Report to the Commission of the European Communities, Directorate General XII—Research, Science, Education, Brussels, November 1977.

11. National Center for Resource Recovery. Materials recovery system: Engineering feasibility study. Washington, D.C., 1972.

12. H. Alter, S. L. Natof, K. L. Woodruff, W. L. Freyberger, and E. L. Michaels. Classification and concentration of municipal solid waste. In Proceedings, Fourth Mineral Waste Utilization Symposium (E. Aleshin, ed.). IIT Res. Inst. and Bureau of Mines, Chicago, 1974, pp. 70-76.

13. E. L. Michaels, K. L. Woodruff, W. L. Freyberger, and H. Alter. Heavy media separation of aluminum from municipal solid waste. Transactions of SME 258: 349 (1975).

14. H. Alter and B. Crawford. Materials processing research: A summary of investigations. Report on Contract 67-01-2944 to the U.S. Environmental Protection Agency, Office of Solid Waste Management. Washington, D.C., 1976.

15. W. Kaminsky, J. .Menzel, and H. Sinn. Recycling of plastics. Conservation and Recycling 1: 91-110 (1976).

16. H. W. Kindler. R-80 lont sich. U: Da Technische Umweltmagazine (1976).

17. F. J. Colon. Het TNO-scheidingssysteem vor huishoudelijk afval. De Ingenieur 86(7): 131-33 (1974).

18. F. J. Colon. Recycling of paper. Conservation and Recycling 1: 129-36 (1976).

19. B. Kreiter. Composition of waste and some possibilities of recovery. Conservation and Recycling 1: 19-30 (1976).

20. F. J. Colon and H. Kruydenberg. The mechanical separation of municipal refuse into useful components and their application as raw materials in industry. In Proceedings, First Recycling World Congress (M. Henstock, ed.). Exhibitions for Industry, Oxted, Surrey, 1978, vol. 4.

21. B. Citron and B. Halén. Automated recovery from domestic refuse. Report STU 73-5182. A. B. Svenska Fläktfabriken, Stockholm, 1978; C. Cederholm. Results from the Fläkt paper fiber recovery system and its application in commercial municipal solid waste plants. In Proceedings, Sixth Mineral Waste Utilization Symposium (E. Aleshin, ed.). IIT Res. Inst. and Bureau of Mines, Chicago, 1978, pp. 188-95.

22. J. N. Gony and F. Clin. B. R. G. M. processes for resource recovery from French urban waste. In Proceedings, Sixth Mineral Waste Utilization Symposium (E. Aleshin, ed.). IIT Res. Inst. and Bureau of Mines, Chicago, 1978, pp. 61-69.

23. F. Clin and J. N. Gony. Household waste sorting systems: Costs and developments in the French economic situation. Report to the Commission of the European Communities, Directorate General XII—Research, Science, Education. December 1977.

24. M. Webb and L. Whalley. Household waste sorting systems. Report to the Commission of the European Communities, Directorate General XII—Research, Science, Education. Brussels, November 1977. (Also issued as report CR 1399 from the Warren Spring Laboratory, U.K. Department of Industry, Stevenage, Herts.)

25. R. Fox. The disposal of urban waste in Italy. Report to the Commission of the European Communities, Directorate General XII—Research, Science, Education. Brussels, August 1977.

26. Sorain-Cecchini S. p. a. Via Bruxelles 53, 00198 Rome.

27. H. W. Kindler and K. Hillekamp. Private communication. October 1975.

28. K. Hillekamp. Neue Technologien zur Rückewinnung von Altpapier aus kommunalen Abfällen, sowei deren Wieder/Weiterwendung innerhalls und auserhalf der Papierindustrie. Wochenblatt für Papierfabrikation, nos. 1-3 (1976).

29. H. W. Kindler. Research into the hygienic qualities of paper recovered by mechanical sorting of municipal waste. Conservation and Recycling 2: 263-68 (1978).

30. H. Alter, K. L. Woodruff, A. Fookson, and B. Rogers. Analysis of newsprint recovered from mixed municipal waste. Resource Recovery and Conservation 2: 79-84 (1976).

Chapter 5

THE TECHNOLOGY OF WASTE-TO-ENERGY CONVERSION
IN THE UNITED STATES AND EUROPE

The major components of municipal solid waste (MSW) in the United States, United Kingdom, and some continental countries are paper, cardboard, plastics, wood, and other combustible materials. Minor components include metals, glass, and other noncombustible materials. The combustible portion can be used either as a fuel or feedstock for conversion to some other form of fuel [1]. These wastes are often overlooked as an alternative to the use of scarce fossil fuels.

Wastes have long been used as fuel. The American Indian used buffalo chips (and presumably other wastes) for heating and cooking before European settlers arrived in North America, a practice probably followed by other early societies and still employed in some developing countries. Heat recovery from incinerators was first practiced before the turn of the century. This method of generating steam and electricity began in Hamburg in 1896 and New York City in 1903 [2]. Most steam units and incinerators were abandoned in the United States after 1945 with the advent of modern (and less expensive) sanitary landfills. Their demise was accelerated during the 1960s by the movement for improved air quality and the general availability of "inexpensive" disposal sites and cheap fuel. By contrast, incineration of wastes for steam and electrical generation generally expanded in Western Europe following 1945 due to the general lack of suitable land disposal sites and other factors. Few steam-generating waste burners exist in the United States, but some 243 such units are installed or under construction in Western Europe (including the United Kingdom) that can process a total of 3,250 tons per hour [3].

Action by the Organization of Petroleum Exporting Countries (OPEC) in 1973 stimulated interest in waste and other alternative fuel sources. The use of municipal wastes as fuel is clearly energy-conservative. For example, as pointed out in Chapter 2, it is estimated that the combustible portion of solid waste in the Standard

58

Metropolitan Statistical Areas (SMSAs) of the United States has the arithmetic equivalent of about 9×10^{17} joules/year (8.5×10^{14} Btu/y). The resource recovery of this portion, and recovery and reuse of metals and glass, could conserve approximately 95×10^6 liters (600,000 barrels) of oil per day [4].

A concern in achieving resource recovery from waste is the amount of energy required to collect and convert it to energy. This subject will be addressed later and analyzed in some detail. However, one analysis suggests that a plant designed to process 590 tons of MSW per day to extract iron and steel, aluminum, other nonferrous metals as a mixture, and glass, and to produce a refuse-derived fuel (to be used as a supplement to coal to generate electricity), could generate 8.5 times more electricity than it would consume [5].

The technology for waste-to-energy conversion in Europe and the United States is evolving along different lines. The major trend in Europe has been to burn unprocessed waste to generate steam (or hot water) for district heating and electric power generation. The newer facilities in the United States mechanically process the waste to upgrade it to a refuse-derived fuel (RDF) for use by itself or as a coal supplement. There is much interest by Europeans in the United States technology and vice versa. One indication is the construction of European-type plants in the United States and RDF plants in Europe.

DEFINITIONS

Before discussing the technology of waste-to-energy conversion, some terms should be defined. The usual measure of fuel value (or energy potential) is the heat of combustion, * or the heat released during complete burning under specified laboratory conditions. The heat of combustion is an inherent property of a substance and is related to its molecular structure. It is expressed in units such as Btu/lb or cal/g; the preferred international unit is joules/kilogram (J/kg). These are merely units of measurement, and a term such as "the Btu content" is jargon, has no meaning, and should not be used.

Since the 1973 OPEC embargo, there has been a tendency to express fuel values in "newspeak" units, such as "barrels of oil equivalent." This is an imprecise practice because it ignores the energy investment and losses (inefficiencies) in processing and use. For example, when MSW is processed to RDF, the fuel value of the RDF is greater than that of the MSW. Achieving this upgrading was one of the

---

*Also called the calorific value.

objectives of the processing. However, the amount or yield of RDF is
less than the amount of MSW processed. The laws of nature dictate
that matter and energy can be neither created nor destroyed. Even
though the RDF is a better fuel (higher heat of combustion than MSW),
there will be less total fuel available. This processing loss, as well
as losses in using the new fuel, can be illustrated by the following
three definitions [4, 6].

The arithmetic equivalence is the heat of combustion, or higher
heating value (HHV), of the waste or some processed form of the waste,
multiplied by a conversion factor to express the HHV in units of barrels
of oil or some other convenient unit for discussion.

The conversion equivalence accounts for energy losses during the
conversion of raw waste to produce fuel or energy. This would be the
HHV of the wastes multiplied by the efficiency of the conversion process—but less the energy input to the conversion process. Another way
of expressing it would be the energy output (as product) of a given process, less the energy required to operate the process, and less the
losses resulting from unrecovered products as compared with input
waste.

The substitution equivalence allows for use of the new fuel as a
substitute for a conventional fuel to produce the same energy product
(such as steam, electricity, heat, etc.). In a given application, the
new fuel may perform with the same, greater, or lower efficiency than
the fuel which it is replacing.

The substitution equivalence is the ratio of the efficiency of use of
a conventional fuel in a given application to the efficiency of use of the
waste-derived fuel in the same application [7]. Obviously, it is the
substitution equivalence that should be used as the basis for policy
planning in choosing among fuels.

Proven technology is another loosely used term. It is rarely clear
if proven in this context means that the technological design and equipment are reliable, that there is a record of consistent operating performance, and that the economics are established. Because of these
ambiguities, the term proven technology is not used in this study.

The term incinerator may also be ambiguous. It has been generally described as a device or installation for the burning and destruc-
tion of solid waste. Recently, modifiers such as "waterwall" or
"steam-generating" have been added to update it by including a method
of resource recovery. Other terms have been proposed, but they seem
cumbersome or imprecise. For clarity, heat recovery incinerator
is used here to mean a unit for burning unprocessed solid wastes to
generate steam or hot water. This process may involve any type of
waterwall or waste heat boiler.

Refuse-derived fuel (RDF) refers to the product derived from the mechanical processing of MSW. RDF can be obtained by several means which will be described later. Strictly speaking, raw MSW is a form of RDF, as is shredded MSW. A case can be made that the liquid or gaseous fuels from chemical or biological treatment of wastes are also forms of RDF. Thus, the term RDF does not apply only to the product of shredding and air classification.

RDF has most often been burned in suspension—blown into a boiler like powdered coal. RDF can also be densified by compression or extrusion into something resembling lump or stoker coal to produce densified refuse-derived fuel (d-RDF).

Various forms of RDF are quite different in appearance and composition, which affects their ability to be conveyed, screened, stored, and so on. For example, Fig. 5.1 is a photograph of shredded MSW. This is material nominally 10 cm (4 in.). For comparison, Fig. 5.2 is a photograph of the air-classified light fraction of the shredded MSW in Fig. 5.1. Fig. 5.3 shows the air-classified light fraction (Fig. 5.2) after it has been reshredded to nominal 2 cm (0.8 in.). Fig. 5.4 shows various forms of experimental d-RDF made from mixed municipal and industrial wastes. The largest piece is 5 cm (2 in.) in diameter.

TYPES OF WASTE-TO-FUEL CONVERSION TECHNOLOGIES

MSW may be used directly as a fuel or processed mechanically to form RDF. In turn, RDF (or similar processed fraction) may be a feedstock for further chemical or biological conversion to another fuel form. This section reviews the processing technology to produce refuse-derived solid fuels. The following section reviews the technology of using these fuels.

The processing of wastes to create solid fuels is illustrated in Fig. 5.5. Each processing step upgrades the fuel by raising the heating value, lowering the ash content, and maybe improving the handleability. The purpose of each additional processing step is to produce a more consistent product by narrowing the range of properties of the fuel delivered to the user. However, generally each additional processing step reduces the yield (mass of fuel recovered) and consumes additional energy. The overall objective of combining processing steps is for the fuel to meet predetermined specifications rather than exhibiting the wide range of properties and heterogeneity of raw waste.

Figure 5.5 displays the interrelationships of the several means of processing wastes and is a useful outline for discussing particular

FIGURE 5.1. Shredded municipal solid waste. A lid from a can, a beer can, and other pieces of metal can be seen. (Photo: H. Alter.)

FIGURE 5.2. Air-classified light fraction. Large pieces of paper are mostly visible. There are other components, such as food waste, sand, grit, and small pieces of metal. (Photo: H. Alter.)

FIGURE 5.3. Secondary shredded air-classified light fraction. This material, as shown, would be the RDF for suspension firing or the feedstock for preparing d-RDF. (Photo: H. Alter.)

FIGURE 5.4. Various forms of densified refuse-derived fuel (d-RDF). The largest piece in the center is 5 cm in diameter. The pieces shown are from various experimental trials (and one production plant) in the United States, England, and France. (Photo: H. Alter.)

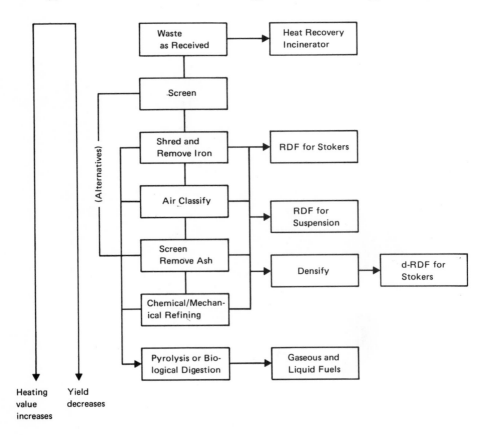

FIGURE 5.5. Types of waste-to-energy processing.

processing schemes. Wet-processing is not shown in Fig. 5.5 but is
discussed below.

The simplest processing of as-received MSW to a fuel may be by
shredding and separation of the magnetic metals (mag. sepn.), represented by

$$\text{Shred} + \text{mag. sepn.} \rightarrow \text{RDF} \qquad (1)$$

This produces a fuel with a high ash content. United States waste contains about 7 percent by weight magnetic metals and perhaps 20 percent

by weight or more noncombustibles. However, the shredding step produces a more homogeneous product.

The next step would be to air-classify the shredded waste, to
separate the "light" materials from the "heavy" materials. The
"lights" consist mostly of paper and plastic film low in noncombustible
or ash content. The "heavies" contain much inorganic material that
is noncombustible. The first processing with air classifying (A/C)
was

$$\text{Shred} + \text{A/C} \to \text{RDF} \tag{2}$$

Magnetic metals may be removed before or after the A/C step. <u>Mag-
netic separation is not shown but is implied in this scheme and all
others below.</u>

The RDF produced by scheme (2) was intended for use in a suspension boiler. Thus, it was shredded small (less than $1\frac{1}{2}$ in. or 3.8
cm). Shredding this small in one step is not considered economical,
so scheme (2) is now practiced as

$$\text{Shred} + \text{A/C} + \text{shred} \to \text{RDF} \tag{3}$$

or

$$\text{Shred} + \text{shred} + \text{A/C} \to \text{RDF} \tag{4}$$

As mentioned earlier, one purpose of air classification is to reduce the ash. However, it has generally been the experience that
schemes (2), (3), and (4) still produce RDF with high ash contents.
Scheme (2), practiced at the St. Louis Union Electric Company demon-
stration, produced RDF containing an average of 26.3 percent ash (dry
weight basis) [8]. The RDF includes removable inorganic fines aris-
ing from dirt, crushed glass, stones, and so on, which are dense but
small enough to "fly" in the air classifier [9]. Much of the removable
inorganic fines can be eliminated by screening, which may be accom-
plished before (10) or after (11) shredding or following air classification (12):

$$\text{Screen} + \text{shred} + \text{A/C} \to \text{RDF} \tag{5}$$

$$\text{Shred} + \text{screen} + \text{A/C} \to \text{RDF} \tag{6}$$

$$\text{Shred} + \text{A/C} + \text{screen} \to \text{RDF} \tag{7}$$

Schemes (5), (6), and (7) show only one stage of shredding. A second
stage can be added. For example, the scheme used at a pilot plant to
produce d-RDF [13] for stoker burning was

$$\text{Shred} + \text{screen} + \text{A/C} + \text{shred} + \text{densify} \rightarrow \text{d-RDF} \qquad (8)$$

The densifying produces a material more easily stored and handled than the RDF fluff product. However, whereas the fluff RDF has been used in both suspension and stoker boilers, the d-RDF can only be used in stoker boilers—unless, of course, it was reshredded.

Further improvements in processing technology produce a proprietary product called "Eco-Fuel II" [14]. This is a sized, dry, free-flowing fuel which is produced by a process resembling the following scheme:

$$\text{Shred} + \text{A/C} + \text{screen} + \left\{ \begin{array}{c} \text{chemical} \\ \text{treatment} \end{array} \right\} + \left\{ \begin{array}{c} \text{hot ball} \\ \text{mill/screen} \end{array} \right\} \rightarrow \text{"Eco-Fuel"}$$

$$(9)$$

Presumably, the chemical treatment catalyzes the air oxidation of the paper in the light fraction (or is itself an oxidizing agent). This step, combined with the heat in the ball mill, causes the cellulose to embrittle (much like baking a piece of paper) so it can be more easily ground.

A form of RDF has been produced by wet-processing. The MSW is first mixed with water and then "shredded" in what may best be described as a giant food blender, the same equipment used to pulp wood chips to make paper. The resultant slurry is passed through hydrocyclones which remove the glass and grit which did not settle out of the slurry. The slurry of paper, plastics, and other combustibles is dewatered to form RDF [15]. Wet-processing is described in scheme (10):

$$\text{Hydropulper} + \text{hydrocyclone} + \text{dewater press} \rightarrow \text{RDF} \qquad (10)$$

Schemes (1) through (10) represent United States technology for mechanically converting waste to a specification fuel. A somewhat different approach has been taken in Europe.

Much European technology is directed toward recovery of usable paper fiber rather than RDF. Conceivably, many of the schemes for paper recovery could be altered slightly to produce RDF. For example, the Fläkt process for paper recovery, described earlier, starts with scheme (6).

Some of the other processes described in Chapter 4 that do not use a shredder may be represented by

$$\text{Screen} + \text{A/C} \rightarrow \text{RDF} \qquad (11a)$$

or by

$$Screen + A/C + shred \rightarrow RDF \qquad (11b)$$

The screen would remove rocks, stones, glass, and other noncombustible materials from the RDF.

Screens were investigated some years ago in England as an alternative to shredders or as a step preceding shredders [16]. The technical and economic advantages of a screen preceding a shredder have been described [10, 17]. There has been an attempt to summarize the mass and energy balances of these various processing schemes as well as for some pyrolysis schemes [18].

Many processes have been proposed for the pyrolysis of MSW to either liquid or solid fuels. Most are little more than "bench scale" or design proposals. A few have been built, even fewer on a scale of more than 1 ton per hour. Many of the pyrolysis plants and proposals were recently reviewed [19, 20].

TYPES OF WASTE-TO-ENERGY USE
TECHNOLOGIES

Raw and processed MSW may be used as fuel, either by themselves or as a supplement to coal, oil, or gas. It may help produce steam or hot water for district and space heating, as well as generate electricity. In some cases, specially designed boilers are used; but existing boilers may be modified. Each of these applications is discussed below.

A. Heat Recovery Incinerators

This technology uses unprocessed MSW as the fuel. Heat recovery incinerators are used in both the United States and Europe but are clearly more popular in Europe. Heat recovery incinerators differ in design [3], but all basically have some sort of grate on which the MSW is spread to burn evenly. Provision is made for overfire and underfire air. The combustion gases are passed by waterwalls and/or arrays of boiler tubes to transfer the heat to water and generate steam. A number of proprietary designs, particularly of the grates and boiler tube arrays, have been proposed and built. They seek to achieve even burn-out and increase heat recovery as well as minimize corrosion and slagging. For example, MSW must be well mixed with air to achieve complete combustion. Incomplete combustion results in chemical reducing conditions that cause corrosion. Three of several different grate designs are shown in Fig. 5.6. Figure 5.7 depicts so-called drum grates in a heat recovery incinerator.

Provision is often made for the addition of supplementary fuel. Once it was believed that MSW would be too wet to ignite or sustain combustion. Generally, this is not the case in the United States and many areas of Western Europe. Operators have learned to mix various sources of waste. For example, dry office wastes are blended with wetter household wastes to avoid the use of supplementary fuel.

Supplementary fuel is used in European installations to provide for constant or increased generation of steam because supplies of MSW often fluctuate. For example, the Munich South heat recovery incinerators use both oil and gas. In 1975, MSW provided only 15 percent of the heat input to the Munich South units IV and V that generate electricity and district heat [21].

There are installations in the United States where a separate fossil fuel boiler is provided at the incinerator site to maintain the steam supply when the heat recovery incinerator is shut down or the MSW supply falls off. Both United States and European practices illustrate that heat recovery incinerators must be reliable since they are energy suppliers and not merely a means of waste disposal.

B. Shredded MSW in Stoker Boilers

Two approaches are employed in the method using shredded MSW in stoker boilers that use RDF produced by scheme (1). The first employs this form of RDF alone; the second mixes it with coal.

The RDF can be used in a traveling grate-stoker unit, as illustrated in Fig. 5.8. This schematic shows a refuse fuel being distributed over a coal spreader through the front wall of the stoker unit. Alternatively, the RDF may be fed through the back wall. Some of the fuel will burn in suspension and the remainder on the grate. This general method of using RDF is practiced in an industrial plant in England [22].

Another method of using the RDF produced by scheme (1) is to burn it by itself (i.e., no coal) in suspension. Because the RDF will contain large pieces of material that will not combust completely in the short time they are in suspension, the boiler for this application is fitted with a grate. Pieces falling through the suspension zone will fall onto the grate and burn. This method of waste-to-energy conversion is practiced in Canada [23]. A schematic of the boiler used is shown in Fig. 5.9.

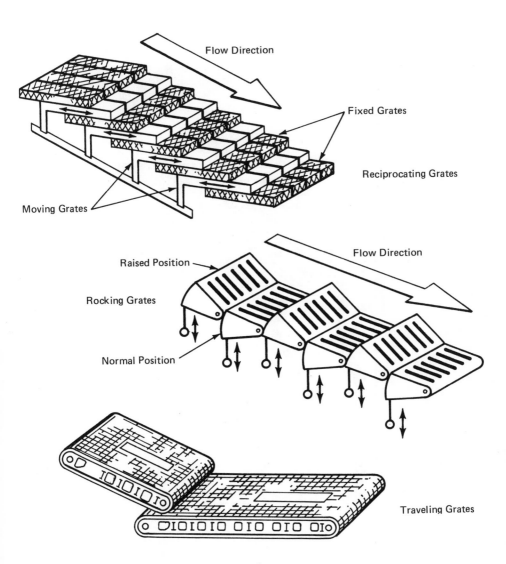

FIGURE 5.6. Types of incinerator grates.

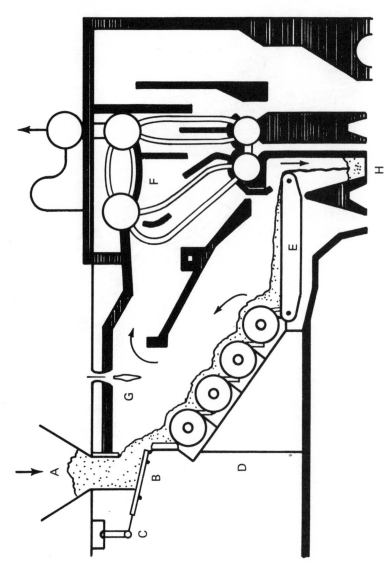

FIGURE 5.7. Diagram of a heat recovery incinerator with drum-type grates. (A) Refuse hopper; (b) refuse feed gate; (C) feed gate drive; (D) rotating grate; (E) traveling grate; (F) steam generator; (G) oil burner; (H) ash and clinker discharge.

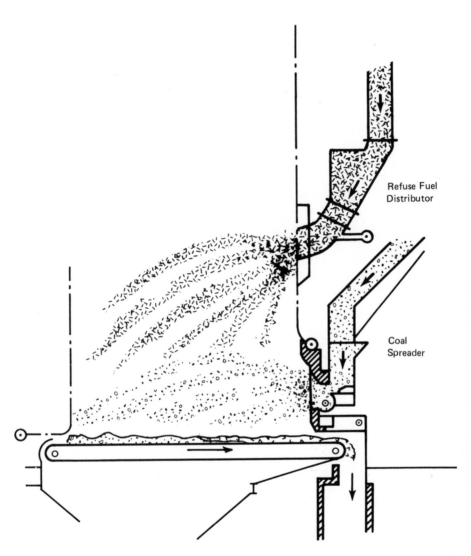

FIGURE 5.8. A method of introducing RDF produced by scheme (1) into a spreader-stoker coal-fired boiler. (Courtesy H. Hollander, Gilbert Associates.)

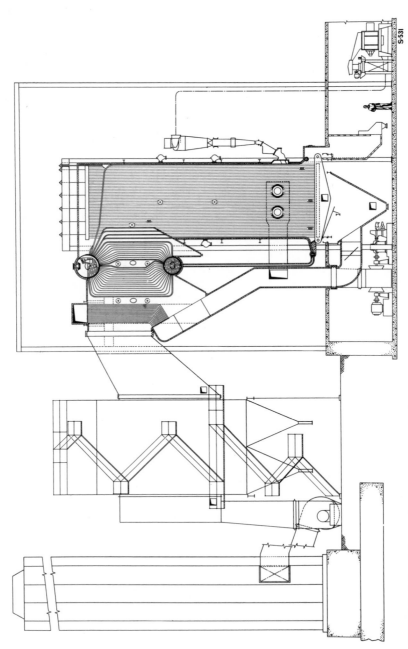

FIGURE 5.9. Schematic cross-section of a boiler used for burning RDF prepared according to scheme (1) by itself. The fuel is charged to burn first in suspension, then on the grate. This particular boiler is in use in Hamilton, Ontario, Canada. (Courtesy the Babcock & Wilcox Company.)

C. RDF in Suspension-Fired Coal Boilers

The principle of suspension firing of pulverized fuels is illustrated in
Fig. 5.10 for a unit where the fuel is fed into the corners. This
approach uses forms of RDF prepared according to schemes (2) through

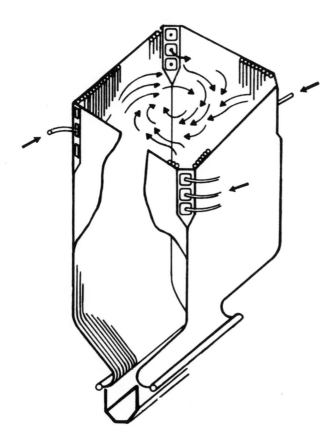

FIGURE 5.10. Schematic representation of suspension firing of pulverized fuels. The fuel is charged to the combustion unit through
nozzles in each of the four corners. The flame pattern impinges on
the boiler tubes lining the walls. The ash falls to the bottom where it
is removed dry or sluiced with quench water. (Courtesy H. Hollander,
Gilbert Associates.)

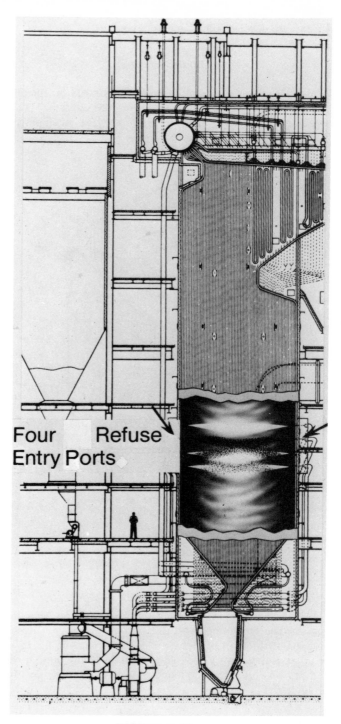

Four   Refuse
Entry Ports

FIGURE 5.11

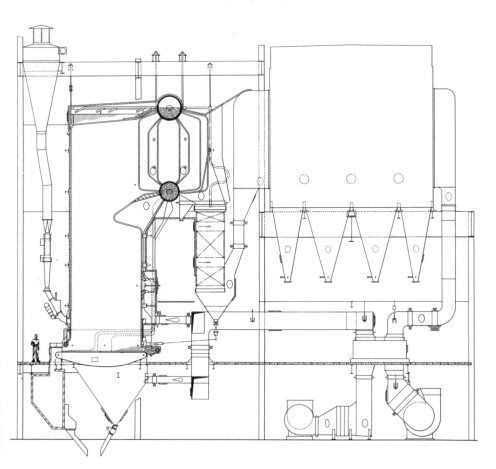

FIGURE 5.12. Schematic representation of a boiler for semisuspension burning of RDF prepared according to schemes (2) through (7). The RDF is fed from the hopper on the left. Whatever does not burn in suspension falls on the grate. The RDF is intended to be 100 percent of the fuel in this boiler. This figure is a schematic of the boilers for Akron, Ohio. (Courtesy the Babcock & Wilcox Company.)

FIGURE 5.11. Diagram of a suspension boiler fitted to burn refuse-derived fuel as a supplement to powdered coal. The RDF entry ports are near the similar ones for powdered coal. (Courtesy Combustion Engineering, Inc.)

(7) as a supplementary fuel with pulverized coal. The coal is reduced in size to about 200 mesh (0.07 mm) and blown into the boiler to burn suspended in the combustion air. The method of feeding the fuel and combustion air may be as shown in Fig. 5.10 or as in other designs where the fuel may be charged through the front wall or elsewhere. This general type of suspension burning is frequently used by electric-generating stations in the United States for large installations. Adding RDF as a supplementary fuel involves replacing (or adding) several of the powdered coal nozzles with RDF nozzles. Figure 5.11 illustrates a large power plant suspension boiler (corner-fired), fitted with RDF nozzles, as might be used by a utility. Another design proposes to add a new upper level of nozzles for RDF to existing units [24].

In technology C, the RDF is shredded to pieces small enough to burn completely in the brief time during fall in the boiler. For example, a specification which has been used is that pieces of RDF have to pass a 3.2-cm ($1\frac{1}{4}$-inch) sieve. However, this may be misleading. A piece of paper passing this sieve size, or even larger, will have its thickness as its minimum dimension. A piece of wood which might be included in the air classifier light fraction, and passing this sieve size, could have as its minimum dimension nearly the full sieve dimension. Initiation and rate of combustion are more related to minimum dimension than to sieve size, thus arguing for separation of wood and other "thick" materials from paper and film plastic. Put differently, if the RDF were prepared to consist only of paper and film plastics, it may be a satisfactory fuel for this application at larger sieve sizes, hence requiring less shredding.

Technology C for waste-to-energy conversion is practiced at several locations in the United States but not in Europe. Examples are given in Chapter 6. The operation of plants using RDF in suspension-fired coal boilers has been documented [24, 25].

D. RDF as the Sole Fuel in Suspension-
    Fired Boilers

As an alternative to technology C, RDF has been proposed as the sole fuel in suspension-fired units. To do so, a grate is added to the bottom of a boiler such as the one shown in Fig. 5.12. This figure is a schematic of the boilers for Akron, Ohio. Similar to the boiler shown in Fig. 5.9, any RDF which does not burn completely in suspension falls onto the grate. This type of boiler, suspension plus bottom grate, is similar to those used for burning wood wastes. RDF produced by schemes (2) through (7) could be used.

E. RDF in Stoker-Fired Boilers

Similar to conversion technology B, RDF prepared by schemes (2) through (7) can be used in a stoker boiler. The practice involves adding the RDF pneumatically through front or back walls so that some burns in suspension and the remainder falls on the traveling grate, as shown in Fig. 5.8.

F. d-RDF in Stoker-Fired Boilers

The d-RDF prepared in scheme (8) can be charged to a stoker boiler and mixed with lump coal, the normal fuel for this type of unit. An advantage of densifying is that the d-RDF can be so mixed and fed through the existing fuel-feed devices, thus avoiding structural changes to the boiler. Furthermore, d-RDF can be stored in simple bunkers, unlike the live-bottom bins needed to contain and convey RDF in fluff form.

G. Use of Wet-Process RDF in Boilers

RDF prepared by a wet process—scheme (10), for example—can be planned to be burned in boiler-types used for bark and similar wet wood wastes. It is proposed to use this form of RDF as the sole fuel, although there is no reason why it could not be used as a supplementary fuel with coal or wood wastes.

H. Codisposal with Sewage Sludge

The codisposal of MSW and sewage sludge in incinerators and heat recovery incinerators has long been practiced. Clearly, enough fuel must be added (RDF in some form) to compensate for the large amount of water present in sludge. Recently, there has been interest in using processed RDF either as a fuel for the sludge in a fluidized bed sludge burner [26] or to be pyrolyzed in a modified multiple-hearth sludge incinerator and the gases evolved used to dispose of the sludge [27].

In summary, any of the conversion technologies can use some form of RDF by itself or as a supplementary fuel with oil, gas, or coal. Not well known in the United States, the practice in Germany has been to use heat recovery incinerators for unprocessed MSW with fossil fuels [21]. Perhaps for this reason, using more highly processed forms of RDF does not appear to be as popular in Europe as in the United States.

THE EFFICIENCY OF WASTE-TO-ENERGY
CONVERSION

Efficiency broadly is the ratio of output to input, and is usually expressed as a percentage. Thus, the municipal planner should know
the efficiency of various systems to obtain the maximum return from
a waste-to-energy plant. Unfortunately, the term efficiency is too
loosely applied to resource recovery systems, though it is unambigu-
ously defined by the laws of thermodynamics. This ambiguity causes
confusion among planners and engineers. For example, Table 5.1
lists values of the "efficiency" of various resource recovery systems
given by two authors who only agree in some cases. Some seven different values of "efficiency" with seven different names were identified in the literature for one system [30]. A rigorous method of computing the efficiency of waste-to-energy systems has been proposed
[30], but is not yet widely practiced.

There is more than one efficiency in describing resource recovery systems. First, there is the conversion efficiency, related to
the conversion equivalence of fuel. This is a measure of how much
energy contained in the input MSW (or its equivalent) is consumed as
it is processed to produce RDF. Second, there is the substitution
efficiency, related to the substitution equivalence defined earlier.
This measures the heat value of the RDF being converted to steam
relative to the heat value of the conventional fuel which is converted to

TABLE 5.1. Energy Recovery Efficiencies

System	Reported efficiency (%)	
	Ref. 28	Ref. 29
Waterwall incinerator	67	57
Solid RDF	66	70
Pyrolysis to oil ("Oxy")	37	37
Pyrolysis to gas ("Purox")	62	63
Pyrolysis to gas ("Torrax")	45	65
Anaerobic digestion to CH_4	25	42

steam in the same unit. (Recall that the substitution equivalence of a
new fuel may be less than or greater than unity, which means a quantity of fossil fuel may be replaced by less than or more than its equivalent, on a heat value basis, of the new fuel.) Finally, there is the
boiler or thermal efficiency, or the measure of the fraction of the heat
value of any fuel which is converted to steam. Because of the laws of
thermodynamics (or the perversity of nature), no conversion system
can be 100 percent efficient. Nor can all of the fuel's heat value be
converted to an equivalent amount of steam.

How these efficiencies apply to the choice of a waste-to-energy
system may be illustrated using a published report of the conversion
efficiency of MSW to solid, liquid, or gaseous fuel; the thermal efficiency when these fuels are used to generate steam; and the substitution efficiency when the steam from either conventional fuel or RDF is
used to generate electricity [31]. The results for five systems are
listed in Table 5.2. The substitution efficiency appears to be low, but
so is the conversion efficiency of fossil fuel to electricity.

In Table 5.2, the conversion efficiency of MSW is not defined.
That is, there is no energy required to convert the MSW to fuel. This
might be used as an argument for using heat recovery incinerators
instead of mechanically processing some other form of RDF. However,
the thermal efficiency of MSW to steam is relatively lower than using
many other forms of RDF. Also, the quality of the steam (its temperature, pressure, and saturation) affects the thermal efficiency of converting it to electricity in a turbine generator. The result is that the
substitution efficiency of converting waste to RDF to generate electricity is slightly higher than using raw MSW to generate electricity.
Some further examples and applications of the various efficiencies of
waste-to-energy conversion have been described [32].

THE PROPERTIES OF REFUSE-
DERIVED FUELS

As indicated earlier, the purpose of processing MSW to some form of
RDF is to upgrade the fuel properties and make them more consistent
than the corresponding properties of as-received MSW. It is intended
to produce what may be termed a "specification fuel." However,
buyers and sellers have yet to agree upon RDF specifications [33].
The processing is also intended to even out normal variations in the
waste's form and composition.

Table 5.3 summarizes some of the reported properties of various
forms of RDF. Included for comparison are fuel properties of as-
received MSW in the United States and Europe. The fuel properties of

TABLE 5.2. Conversion and Substitution Efficiencies

System	Conversion Efficiencies (%)		Substitution Efficiencies (%)
	To fuel	To steam	To produce electricity
MSW, heat recovery incineration combustion at 100% excess air	b	60.5[c]	21.0
RDF, combustion at 30% excess air	85.2	54.0[d]	23.0
Pyrolysis to liquid fuel "Oxy" process combustion at 10% excess air	52.8	32.5[d]	13.8
Pyrolysis to gas "Purox" process combustion at 15% excess air	77.7	51.6[d]	21.9
Anaerobic digestion to 95/5, CH_4/CO_2 combustion at 5% excess air	41.7	22.8[d]	9.9

[a] Based on waste of 10.5 MJ/kg (4500 Btu/lb) substituted for coal, gas, or oil in the same application.
[b] Undefined.
[c] Steam at 4.50 MPa (650 psi), 470° C (875° F). Turbine heat rate 2.86 (9750 Btu/kWh).
[d] Steam at 12.4 MPa (1800 psi), 540° C (1000° F). Turbine heat rate 3.4 (10,000 Btu/kWh).
Source: Ref. 31.

TABLE 5.3. Typical or Representative Ranges of Fuel Properties of RDF, Dry-Weight Basis

Scheme used for RDF preparation	Heating Value MJ/kg (Btu/lb)	C (wt. %)	H (wt. %)	N (wt. %)	S (wt. %)	Cl[a] (wt. %)	Ash (wt. %)	Moisture as received (wt. %)
MSW–U.S.	7.0-11.6 (3000-5000)	35.3	4.4	0.42	0.20	0.66	—	20-40
MSW–Germany	4.2-12.5 (1800-6700)	—	—	—	—	—	—	—
(1)	15.5 (6700)	—	—	—	0.14	0.43	28.6	30
(2)	16.0 (6900)	38	—	0.6	0.15	0.47	25.5	3-63 (27.5 avg.)
(4)	16.9 (7270)	40.1	6.2	0.54	0.56	0.31	22.6	23.0
(8)[b]	15.5 (6680 ± 9%)	45.6 ±9.5%	6.3 ±23.7%	0.9 ±87.5%	0.3	0.2	16-24	17-25 (19.0 avg.)
(9)	18.0 (7740)	—	—	—	—	—	9.4	2.0

[a]Total Cl, most of which is water-soluble and presumed to be NaCl.
[b]Analytical results are for the RDF before densifying.
Source: Averaged from Refs.

any form of RDF must be related to the composition of the starting
material. For example, a waste rich in plastic film is likely to produce an RDF of high heat value.

Table 5.3 summarizes data from several sources and lists what
may be considered average or typical values (ranges). However, even
when a single value for a property is listed, it is not exact. MSW, and
to a lesser extent RDF derived from it, varies in composition and fuel
properties. For example, wide ranges have been reported for RDF
produced by scheme (2) [8]. Since standard methods of analyzing RDF
do not yet exist, as they do for coals and other fuels, different laboratories obtain contrasting results with the same material. This accounts for the range of values listed in Table 5.3 for RDF produced by
scheme (8). The ranges of heating values listed for European wastes
were used in the design of various heat recovery incinerators and are
not measurements of MSW samples [3].

The value for the chlorine content in Table 5.3 is for total chlorine.
Klumb's [34] contention that much of the chlorine in RDF is in a water-
soluble form, likely common table salt (sodium chloride), has been
confirmed by others [35]. The fact that most of the chlorine is in this
form may explain why there are no reported observations of chloride-
induced stress corrosion from burning RDF with coal. However,
sodium chloride cannot be ruled out as a possible source of chlorine
in the flue gases and hence a source of possible corrosion [37]. Knowledge of the chemical composition of various forms of RDF can be used
to maintain low corrosion when burning it with coal [37].

It is unlikely that the plastic polyvinyl chloride (PVC) is responsible for much corrosion, at least in the United States. Plastics accounted for only about one-third of the total chlorine in one city's
waste [35]. Furthermore, if RDF is produced by schemes involving
air classification, most (if not all) of the PVC is likely to be included
in the heavy fraction, not the light or fuel fraction. In the United States,
PVC is generally used to manufacture heavier extruded or molded
items, rather than light film.

Further increases in plastics waste volumes will likely not include large proportions of types containing chlorine. Additional plastic materials in the waste stream will cause the MSW's heating value
to increase, but the heating value of RDF will not necessarily rise.
Many of the waste conversion technologies utilize an air classifier
which should separate heavy plastic items from the light or fuel fraction. Table 5.3 is a starting point for planning and suggests there is
no one set of values for the fuel properties of MSW or RDF.

NOTES AND REFERENCES

1. The growing worldwide scarcity of petroleum and natural gas is generally accepted. Coal supplies are ample for centuries to come. However, the extraction of coal (or other minerals) may have large and negative effects on the environment and society. For example, the extraction of new mineral deposits can result in pollution, extensive changes in land use, increased energy consumption, and instability of human settlements. See D. B. Brooks and P. W. Andrews. Mineral resources, economic growth, and world population. Science 185: 13-20 (1974).

2. R. Fenton. Current trends in municipal solid waste disposal in New York City. Resource Recovery and Conservation 1: 167-76 (1975); K. Feindler. Paper prepared for the Energy Bureau, Inc. New York, 1976.

3. Resource Planning Associates. European waste-to-energy systems: An overview. Report CONS-2103-6, UC 95e. Dept. of Energy, Washington, D.C., 1977. A unit is defined as a facility built at one time in a single location, contrasted to a plant which is the building in which one or more units may be installed.

4. H. Alter. Energy conservation and fuel production by processing solid waste. Environmental Conservation 4: 11-19 (1977).

5. H. P. Sheng and H. Alter. Energy recovery from municipal solid waste and method of comparing refuse-derived fuels. Resource Recovery and Conservation 1: 85-94 (1975).

6. H. Alter. Energy equivalents. Science 189: 175 (1975).

7. R. C. Baillie and D. M. Doner. Evaluation of energy substitution equivalents. Resource Recovery and Conservation 1: 188-91 (1975).

8. Edison Electric Institute. Case studies on solid waste utilization projects. New York, 1977, p. 91.

9. D. N. Fan. On the air classified light fraction of shredded municipal solid waste: I. Composition and Physical Characteristics. Resource Recovery and Conservation 1: 141-50 (1975).

10. National Center for Resource Recovery. New Orleans implementation study. Washington, D.C., 1977.

11. R. M. Hill. Effective separation of shredded municipal solid waste by elutritation. Waste Age 5 (10): 350-3 (1974).

12. G. Savage and G. J. Trezek. Screening of shredded municipal solid waste. Compost Science 17: 7-10 (1976).

13. H. Alter and J. M. Arnold. Preparation of densified refuse-derived fuel on a pilot scale. In Proceedings, Sixth Mineral Waste Utilization Symposium (E. Aleshin, ed.). IIT Res. Inst. and Bureau of Mines, Chicago, 1978.

14. R. M. Beningson, K. J. Roger, T. J. Lamb, and R. M. Nadkarni.
     Production of Eco-Fuel-II from municipal solid waste.   In:
     Proceedings, First International Conference, Conversion of
     Refuse to Energy. ASME, New York, 1975, pp. 14-21.

15. General Electric Company. Solid waste management technology
     assessment. Van Nostrand Reinhold, New York, 1975, pp. 44-58.

16. Early work was performed at the Greater London Council, Public
     Health Engineering Department in 1964. See J. L. Warren. The
     use of a rotating screen as a means of grading crude refuse for
     pulverization and compression. Resource Recovery and Conser-
     vation 3: 97-112 (1978).

17. K. L. Woodruff. Preprocessing of municipal solid waste for
     resource recovery with a trommel. Transactions, Society of
     Mining Engineers 260: 201-204 (1976).

18. Bechtel Corporation. Fuels from municipal refuse for utilities:
     Technology assessment. Report EPRI 261-1 for the Electric
     Power Institute, San Francisco, 1975.

19. B. Baum and C. H. Parker. Solid waste disposal, vol. 2.
     Reuse/recycle and pyrolysis. Ann Arbor Science, Ann Arbor,
     Mich., 1973.

20. J. J. Jones, R. C. Phillips, S. Takaoka, and F. M. Lewis.
     Pyrolysis, thermal gasification, and liquefaction of solid wastes
     and residues: Worldwide status of processes (as of fall, 1977).
     In Proceedings, National Solid Waste Processing Conference.
     ASME, New York, 1978.

21. Resource Planning Associates. European waste-to-energy sys-
     tems case study of Munich. Report HCP/M103-004. U.S.
     Department of Energy, Washington, D.C., 1977, pp. 48-49.

22. Imperial Metals Industries (Kynoch) Ltd., P. O. Box 216, Wilton,
     Birmingham, B6 7BA, England.

23. Ref. 18, pp. 3-7.

24. Ref. 8, pp. 66-96.

25. D. E. Fiscus, P. G. Forman, and J. D. Kilgore. Refuse proc-
     essing plant equipment, facilities, and environmental considera-
     tions at St. Louis-Union Electric refuse fuel project. In Proceed-
     ings of the 1976 National Waste Processing Conference. ASME,
     New York, 1976, pp. 373-384; Bottom ash generation in a coal-
     fired power plant when refuse-derived supplementary fuel is used.
     In ibid., pp. 481-511. See also J. C. Even. Evaluation of the
     Ames solid waste recovery system, Part I; Summary of environ-
     mental emissions: Equipment, facilities, and economic evalua-
     tions. Report EPA 600/2-77-205. Municipal Environmental
     Research Laboratory, Environmental Protection Agency, Cincinnati,
     1977.

26. The Black Clawson plant in Franklin, Ohio, used RDF from a process similar to scheme (10) in a fluidized bed sludge burner to dispose of the sludge. There was no heat recovery. A plant in Duluth, Minnesota, uses RDF from a scheme similar to (2) as the fuel to dispose of sludge with heat recovery and power generation for in-plant use.

27. R. B. Sieger and P. M. Maroney. Incineration: Pyrolysis of wastewater treatment plant sludges. Prepared for the Environmental Protection Agency Technology Design Seminar for Sludge Treatment and Disposal. Brown and Caldwell, Walnut Creek, Calif., 1977; B. D. Braiken, J. R. Coe, and T. D. Allen. Full scale testing of energy production from solid waste. Brown and Caldwell, Walnut Creek, Calif., n.d.

28. J. B. Benziger. Resource recovery technology for urban decision makers. Urban Technology Center, Columbia University, New York, 1976.

29. A. Poole. The potential for recovery from organic wastes. In The energy conservation papers (R. H. Williams, ed.). Ballinger, Cambridge, Mass., 1975.

30. R. C. Bailie and D. M. Doner. Evaluation of the efficiency of energy resource recovery systems. Resource Recovery and Conservation 1: 177-187 (1975).

31. R. S. Hecklinger. The relative value of energy derived from municipal refuse. In Proceedings of the 1976 National Waste Processing Conference. ASME, New York, 1976, pp. 133-140.

32. H. Alter. Energy recovery from solid waste: Looking through a dark furnace slowly. In Present status and research needs in energy recovery from solid waste (R. A. Matula, ed.). ASME, New York, 1977, pp. 9-22.

33. The development of RDF specifications and the related problem of developing standard test methods is an ongoing activity of Committee E-38 of the American Society for Testing and Materials, 1916 Race Street, Philadelphia, Pennsylvania 19103.

34. D. L. Klumb. Union Electric Company's solid waste utilization systems. Resource Recovery and Conservation 1: 225-234 (1976).

35. H. Alter, G. Ingle, and E. R. Kaiser. Chemical analyses of the organic portions of household refuse: The effects of certain elements on incineration and resource recovery. Solid Wastes Management (Inst. of Solid Wastes Management, England) 64(12): 706-712 (1974).

36. H. Alter. Resource recovery from a chemical viewpoint. In Materials and national policy. American Chemical Society, Washington, D.C., 1979, pp. 35-44.

37. D. A. Vaughn, H. H. Krause, and W. K. Boyd. Fireside corro-
     sion in municipal incinerators versus refuse composition. Mate-
     rials Performance 14: 16-18 (1975); ibid., Corrosion deposits
     from combustion of solid waste, part IV: Combined firing of
     refuse and coal. ASME Paper No. 75-WA/CD-4. ASME, New
     York, 1975.

Chapter 6

EXAMPLES OF WASTE-TO-ENERGY CONVERSION
AND USE TECHNOLOGIES

This chapter is a tabular compilation of examples of waste-to-energy
conversion and use technologies in the United States and Europe. Most
are for municipal solid waste; some for industrial waste. The lists
and information are intended to be illustrative and not necessarily
complete.

Wherever possible, the examples listed are related back to the
notation system in Chapter 5 for conversion technologies (1) through
(11) and use technologies A through H.

In most instances, references are given to direct readers to
sources of additional information. However, this was not always
possible since some data came from personal contacts and unpublished
documents.

Many plants were in various stages of shakedown and development
when the information was assembled. Details of plant operation and
equipment specifications may change as the processes and plants mature.
Examples of the combination of different processing and use technologies are given in Table 6.1.

In the Netherlands, 30 percent of the total quantity of the municipal
solid waste is incinerated in eleven plants. Five of these are equipped
for energy recovery, most in the form of electricity. As a result, 24
percent of the Dutch waste is used for energy recovery.

Table 6.2 summarizes the types of refuse incineration plants operating in West Germany according to their size and outputs. In Germany,
more than 90 percent of the wastes are collected by regular municipal
or private services. The distribution among the three common disposal methods, landfill, incineration, and composting, is shown in
Table 6.3. Note the majority of waste is landfilled. The comparable
figure for Sweden is that 30 percent of the waste is incinerated and the
remainder landfilled.

89

TABLE 6.1. Waste-to-Energy Plants, United States and Europe (Illustrative Listing)

Conversion technology	Use technology	Plant location	Plant capacity (Mg/working day)[a]	Boiler capacity (Mg/h of steam)	Steam properties	Reference	Status as of July 1979
MSW, as received	A	Saugus, Mass	1089	2 units, 84 each	4.5 MPa, 440°C	1	Operating
		Nashville, Tenn.	363	50	2.5 MPa, 300°C	2	Operating
		Harrisburg, Pa.	653	45.4	1.8 MPa, 240°C	8	Operating
		Chicago, Ill.	1451	4 units, 45 each	1.7 MPa, 240°C	8	Operating
		West Germany	48–960	—	—	3	Operating
		Munich, West Germany[b]	2 units, 600 each; 3 units, 960 each	2 units, 100 each; 3 units, 365 each	18.4 MPa, 540°C	4	Operating
		Kørsor, Denmark	(2 Mg/h)	4.9	0.6 MPa, 120°C (hot water)	5	Operating
(1)	B	Albany, N.Y.	544	2 units, 45 each	1.7 MPa, 232°C	—	Start-up in 1980
		Hamilton, Ont.	454	2 units, 45 each	1.7 MPa, 230°C	—	Operating, no export of steam
		Kynoch, England	250	—	—	7	Operating
		Pueblo, Colo.—Zupan Enterprises	72	9	1.0 MPa, 135°C	—	Operating
(1)	E	Pontiac, Mich.—GM Truck & Coach Div.	—	68	13.8 MPa, 195°C	—	Operating
(2)	C	Rochester, N.Y.—Eastman Kodak Co.	163 waste 122 sludge	61.3	2.7 MPa, 288°C	9	Operating

(2) or (8)	H	California— Contra Costa Co.	—	72.6	4.2 MPa, 440° C	—	Planning
(3)	D	Akron, Ohio— City of Akron	907	3 units, 57 each	4.0 MPa, 248° C	—	Shakedown
(3)	D	Niagara Falls, N.Y.— Hooker Chemical Co.	1996	2 units, 136 each	8.6 MPa, 400° C	—	Under construction, operational 1980
(6)	B	Madison, Wisc.	181	2 units, 181 each	8.6 MPa, 510° C	—	Processing plant: Operating; power plant: shakedown
(7)	C	Monroe County, N.Y.— Rochester Gas & Elec.	1814	40 and 2 units 62 each	10.4 MPa, 540° C	—	Shakedown
(8)	F	Bridgeport, Conn.— United Illuminating Co.	1633	2 units, 284 and 680	10.5 MPa, 540° C	—	Operational 1980
(10)	G	Hempstead, N.Y.	1814	2 units, 181 each	4.6 MPa, 400° C	—	Shakedown

[a]Mg = 1000 kg = 1 ton.

[b]Units North 1A, 1B, and II burn coal and waste. Units 1A and 1B are designed for up to 40 percent waste; unit II for 20 percent waste. Units South IV and V burn gas and waste and are designed for up to 20 percent waste.

[c]Shred (flail) + screen + shred + A/C → RDF.

TABLE 6.2. Classification of Refuse Incineration Plants in West Germany

Size of plant (throughput)	No. with no heat recovery	No. with heating and/or industrial steam	No. with power generation	No. with power and heating
1–3 tons/h	1	2	—	—
3–10	4	5	2	—
10–30	—	5	2	4
> 30	—	2	9	7

Source: Ref. 10. Reproduced by permission of The Institute of Solid Wastes Management (U. K.).

TABLE 6.3. Disposal of Wastes in West Germany

Disposal method	No. of plants	% of total	Annual tonnage
Landfill	—	75.9	13×10^6
Incineration	39[a]	21.7	4.5
Composting	19	2.4	9.5

[a]In 1975.
Source: Ref. 11.

NOTES AND REFERENCES

1. J. T. Kane. Boston area refuse turns into steam for G.E. plant. Professional Engineer 45: 14-17 (1975).
2. C. G. Bozeka. Nashville incinerator performance tests. In Proceedings of the 1976 National Waste Processing Conference. ASME, New York, 1976, pp. 215-27.
3. L. Barniske. Energy recovery from solid wastes in the Federal Republic of Germany: State of the art. Paper presented at the Symposium on Solid Waste Conversion to Energy, Hamburg, 1977.
4. Resource Planning Associates. European waste-to-energy systems: Case study of Munich: Munich IA & IB, II, Munich South IV & V. Report HCP/M2103-0004, U.S. Dept. of Energy, Washington, D.C., 1977.
5. Resource Planning Associates. European waste-to-energy systems: Case study of Kørsor, Denmark. Report HCP/M-2103-0003, U.S. Dept. of Energy, Washington, D.C., 1977.
6. Resource Planning Associates. European waste-to-energy systems: Overview. Report CONS-2103-6, U.S. Dept. of Energy, Washington, D.C., 1977.
7. Imperial Metal Industries (Kynoch) Ltd., P.O. Box 216, Wilton, Birmingham, B6 7BA, England.
8. G. Stabenow. The Chicago Northwest and Harrisburg incinerators. In Proceedings of the 1976 National Waste Processing Conference. ASME, New York, 1976, pp. 81-96.
9. R. L. Merle, M. C. Young, G. R. Love. Design and operation of a suspension fired industrial solid waste disposal system for Kodak Park. In Proceedings of the 1976 National Waste Processing Conference. ASME, New York, 1976, pp. 151-62.
10. L. Barniske. Energy from waste utilization in the Federal Republic of Germany. Solid Wastes 68: 64-82 (1978).
11. CREST study of household waste sorting systems: Final report to Commission of the European Communities. Umweltbundesamt, Berlin, 1978.

Chapter 7

DECISION MAKING FOR WASTE-TO-ENERGY
SYSTEMS PLANNING AND IMPLEMENTATION

This chapter presents a method of calculating the net cost of a waste-to-energy project so that it is directly comparable to current waste disposal costs. The method consists of first establishing the markets for recovered materials and fuels, using the specifications for these markets as the basis for choosing a system, and computing the revenues necessary to make the project economically sound, compared to current costs for disposal. Thus, the cost of waste-to-energy conversion is computed and compared to the competition to determine whether waste-to-energy conversion should be implemented. The essence of the method is fitting the technology to the market, rather than choosing a system because it is in vogue, or because there is a similar plant operating nearby, or because the technology is reported to be successful in another country.

The previous discussions of materials and energy recovery from wastes in the United States and Europe purposefully omitted any mention of cost. Costs in Europe are generally not directly comparable to what the same system or plant may cost in the United States. Further, costs are often reported in ways that make it impossible to relate them to other locations. For example, the plant costs are often given without mention of whether they apply to just the plant, or include necessary modifications to the fuel customer's facility, or perhaps include full project costs such as financing. In short, there is a large difference between plant construction costs and total project costs.

OUTLINE OF THE METHOD

The first step is to identify local markets for recovered materials and fuel. Identify, in this use, means securing a binding letter of intent (LoI) to either bid for or purchase the recovered products to specification. The methods of marketing recovered products and securing such letters of intent, are described in Chapter 8. The second step is to estimate the amount of recovered products which can or might be produced. The third is to obtain a preliminary engineering design and associated cost estimate. The design must be for a plant which will produce the energy product and materials to the specifications determined in the first step. The cost estimate at this point is not precise because of uncertainties in plant construction costs, bid structures, and so on, and should not be considered as the actual budget. Fourth, the costs and revenues must be analyzed and presented in a manner that facilitates public policy decisions. This general approach to decision making has been described and was used in planning a resource recovery system for one city [1, 2].

DEFINITIONS

A waste-to-energy system is the mechanical, chemical, or biological processing of municipal solid waste (MSW), or its processed fraction, to produce a fuel, steam, or electrical energy in a form that fulfills a customer's specifications.

Materials recovery is the mechanical processing of waste, or its fractions, to recover materials in a form suitable for sale. Materials recovery is stressed because the revenue from materials recovery can be a substantial portion of the total revenue of a resource recovery plant. In fact, it is probably not reasonable to contemplate energy recovery without some materials recovery. The marginal revenues from materials recovery far exceed marginal costs in most cases. This is especially true where MSW is to be processed for energy recovery.

An economic model is a systematic way of presenting costs and revenues that enables public decision markers to judge the feasibility of instituting a waste-to-energy system.

A distinction should be made between plant costs and project costs. The former apply to the waste-to-energy processing facility and may or may not include ancillary equipment at the energy customer's plant, such as fuel storage and handling facilities. A project cost may include items such as financing charges, upgrading existing steam lines, new rolling stock, and back-charging for early planning and consultant fees. Plant costs are referred to in the following analyses because there is sufficient experience to predict them. They are bounded and likely to be nearly equal for the same type of plant in different communities. Project costs may be unbounded and will reflect the contrasting needs of communities. Fortunately, there is now sufficient experience to predict some plant costs. In the following analyses, all cost items are identified to differentiate between plant and project costs.

Early analyses are for planning purposes, to decide whether or not to proceed to the next step, and should be undertaken cautiously. Unfortunately, the first planning estimates are sometimes confused with budget estimates and ultimately with budget submissions. When this happens the budget is never enough to pay for the project. Budget estimates must be prepared with great care by building on planning estimates and other, more detailed information.

ESTIMATING THE YIELD OF A WASTE-TO-ENERGY PLANT

To estimate the yield of a waste-to-energy plant, three types of information are required: the amount of waste the community generates, the fraction that can be collected and aggregated for processing, and the composition of the waste in terms of recoverable portions.

The Amount of Waste Generated

The best way of determining the amount of waste generated, or actually collected, is to weigh collection trucks. In the absence of such historical data, estimates have to be used. One method is to determine the volume of waste collected and multiply it by an "average" density. However, the volume of waste collected is often confused with the number of trucks tipping, and some of these may not be full. Also, the "average" density is never known with precision.

A second method is to use published guidelines for average per caput generation. Such averages do not take into account the particular mix of household, industrial, and commercial wastes in different communities. Some jurisdictions have overestimated waste generation using these factors and consequently overbuilt processing capacity.

There are ways to avoid overestimation. First, the period of several years between planning and implementation can be used to obtain reliable data. This can be done by installing a truck scale. Second, estimates of future population and waste generation should be viewed with caution. (These points are discussed in Chapter 3.)

A resource recovery plant need not be planned to process all waste. All recovery systems require a landfill because not all items in the waste can be recovered in a marketable form. It can also serve as a backup disposal facility when the recovery plant is inoperative.

The Amount of Waste Collected

Not all waste is collected and aggregated in a manner that permits processing for resource recovery. For examples, farm communities may dispose of their own waste, private collectors may operate landfills, and private industries may dispose of their waste. Although a community may have the public health responsibility for safe disposal of solid waste, it may not "own" the wastes. Legislating that all waste must be disposed of in a public facility (for example, on the basis of state police powers for maintenance of the public health) causes conflicts because in a sense it expropriates the property of private waste-management firms.

Obviously, a community should not plan a recovery facility for all of its waste if some of it is disposed of by private firms. To do so portends financial difficulties for the facility. The first "law" of resource recovery is that you have to own the waste and plan to process only the portion that will be delivered [3].

The Composition of the Waste

When planning a recovery facility, the composition of the input waste must be viewed categorically, in terms of recoverable materials and fuels. This is often different from past surveys and analyses of total waste composition. For example, the copper in occasional pieces of wire is not recoverable; paper and plastic packaging is not recoverable as such, especially if the plastic is laminated to the paper; light aluminum foil is likely to be combined with the fuel fraction rather than with recoverable aluminum cans.

The average composition of household waste has been reported [4], and this source generally describes the waste in communities but with notable exceptions. For example, few newspapers are as heavy as the New York Times or Washington Post. Also, reading habits vary greatly across the country, and the amount of newsprint in municipal solid waste may range from 5 to 15 percent by weight. A second exception is the amount of recoverable aluminum in waste. Aluminum beer and beverage cans are not marketed everywhere. If many such cans are sold in a community, there is likely enough recoverable aluminum to make its recovery worthwhile. The total amount of aluminum will be small, perhaps of the order of 1 percent, but the revenue from even half this amount is substantial and can help pay for other recovery processes.

Municipal solid waste arises from three sources: household, mixed commercial and industrial, and construction and demolition.

Obviously, waste from the third source should not be mixed with the portion destined for recovery processing. The composition of waste delivered to a recovery plant will depend greatly on the relative mix of the other two sources. Commercial and light industrial wastes vary greatly in composition; heavy industrial waste is usually processed and disposed of by the private sector and never enters the municipal waste stream. Commercial and light industrial waste may contain far less metal and glass and more combustible material than household waste.

The composition of waste in a community can be determined by sampling and handpicking. However, there is some question if this is worthwhile. Sampling obtains the data for just the day and the community sampled. Waste quantity and composition vary with such factors as the time of year, day, or week, the seasons, the weather, and general economic conditions. The quantity and composition may be affected by major marketing changes and innovations—for examples, a newspaper may go out of business, a popular brand of beer may switch to aluminum cans, and so forth. Sampling provides numbers to be written on contracts but is usually too costly to provide reliable data for planning and design.

Sampling is generally so expensive and time consuming that other methods of estimating waste composition have been used. Published average composition can be used, perhaps after some adjustment for local variations in the quantity of newspapers, aluminum packaging, and perhaps yard waste (which is related to climate). The composition and quantity of waste from certain types of commercial activities may be estimated from published guides [5]. The composition must be multiplied by the estimated recovery rates. That is, any recovery process is not perfect and some material will be missed. Such information has not been assembled in one place. Thus, several must be consulted to estimate recovery rates [6-8].*

Finally, it is possible to write a contract for the operation of a recovery plant based on a target composition and then adjusting disposal fees according to experience [6-9]. The fees may be raised or

---

*
The recommendation to seek an alternative to sampling is made with some hesitation. Large variations in composition do exist, city to city. For example, in the United Kingdom the variation in composition of the waste from the national average in one community was so large, and the combustible content so low, that a scheme for recovering RDF had to be hastily withdrawn. Practice in this community is to use coal-fired, individual room heaters, rather than central heating, which might account for the low content of combustible material. However, this experience does illustrate a necessary caution in planning resource recovery.

lowered according to what is delivered and recovered compared to the original target figures in the contract. However, the amount of waste delivered to the recovery facility should be assured so that an increase in disposal fees does not cause less waste to be delivered.

ENGINEERING DESIGN AND PROBABLE COST

The information pertaining to quantity, composition, specifications, and markets is the basis of plant design. These are assigned to an engineering team so that the "first cost" of a plant (or project) can be estimated for the economic model.

There is now enough experience so that some preliminary estimates of plant costs can be made, at least in terms such as dollars per daily ton of operating capacity. Some summaries of such costs have been published [10-12].

THE ECONOMIC DECISION MODEL

The economic decision model used here is based on the concept of the indifference value [1, 2]. It computes what is required to make resource recovery cost the same as current disposal methods. If both are equal, the community should be economically indifferent as to which is used. Obviously, if resource recovery costs less, the economic decision is to implement recovery.

The indifference model keps the tip fee (also called the dump or disposal fee) constant for disposal and recovery. This contrasts with most other calculations of resource recovery costs, which compute a new tip fee for recovery [13].

The starting point for the indifference model is knowing the costs and revenues of the proposed plant. The elements and categories of both are respectively listed in Table 7.1 and Table 7.2. Each will be discussed and the values entered into a generalized statement, or table, of costs and revenues to compute the "cost" of a proposed project.

An important element of cost is the capital cost of the plant or project. The amortized capital cost depends on the method chosen to recapture and pay back the capital investment. If the plant is a public sector investment, the plant is likely to be built as a 100 percent debt investment, repaid through some sort of bonding. The type of bonding will determine the interest rate and finance cost. Some different types of bonding available to the public sector are described in Table 7.3. Perhaps there is a future trend for revenue bond financing [12].

TABLE 7.1. Elements of Cost

Amortized capital cost

Operating cost

    Salaries and wages

    Fringe benefits

    Insurance

    Taxes

    Fuel and other utilities

    Transportation

    Maintenance and supplies

Disposal of unrecovered fraction and residue

TABLE 7.2. Elements of Revenue

Tip fees

Sale of received materials

Sale of energy product

If the recovery plant is a private investment, equity may reduce 100 percent debt financing. However, all-debt financing may be possible through some sort of pollution control bond or other debt instrument that resembles public sector financing. In such cases, the interest on the bonds is free of federal income taxes [14]. Private financing methods are also described in Table 7.3.

The indifference model may be illustrated by the following hypothetical case. A community is able to gather for processing 1000 tons per day of mixed municipal solid waste as described in Table 7.4. The recovery objectives are refuse-derived fuel (RDF) for burning in suspension by the local utility and the recovery of magnetic metals,

TABLE 7.3. Possible Financing Methods for Resource Recovery Plants

Sector to which available	Type of financing	Likely interest rate (1978)	Features
Public	General obligation bonds	$5\frac{1}{2}\%$	Guaranteed by faith and credit of taxable base of the community; adds to community's general indebtedness.
	Revenue bonds	7–8%	Guaranteed by project revenues; requires detail analyses and projection of project's success and ability to sell products. There may be some pressure from the financial community to guarantee the supply of refuse (and concomitantly the revenue from the tipping fee). This may prove to be awkward in some locations.
Private	Industrial development or revenue bonds	$7\frac{1}{2}$–$8\frac{1}{2}\%$	
	Pollution control bonds	$7\frac{1}{2}$–$8\frac{1}{2}\%$	Subject to approval by the Internal Revenue Service.
	Chattel or corporate bonds[a]	8–12%	

[a] Interest subject to federal and local income taxes; other instruments listed are not.

TABLE 7.4. Assumed Characteristics of Community Waste, 1000 (U.S.) Tons per Day, 300 Days per Year

Component in waste	Composition[a] × wt. %	Fraction recoverable[b] =	Percentage available × 300,000 =	U.S. tons per year for sale or disposal
Iron and steel as can stock	6	0.95	5.7	17,100
Heavy miscellaneous iron	1	0.90	0.9	2,700
Aluminum cans and foil	0.5	0.70	0.35	1,050
Mixed nonferrous metals	0.2	0.50	0.10	300
Glassy aggregate	13	0.65	8.5	25,350
RDF[c]	80	0.75	60	180,000
Unrecovered	—	—	24.45	73,500
			Total:	300,000

[a] Estimated from Ref. 4.
[b] Estimated from several sources. See Refs. 6, 7.
[c] Recovered for example by scheme (5), Chapter 5, which removes much of the ash in the RDF.

aluminum, and a mixture of nonferrous metals. Furthermore, the community has decided that a glass-rich mixture will be recovered and given away for construction purposes, rather than sold or landfilled. Thus, it neither incurs a cost of disposal nor generates revenue.

There will be an unrecovered fraction which must be disposed of in a landfill. It is another "law" of resource recovery that there will always be such a residue [3]. The proposed plant is presumed to operate 300 days per year, which is equivalent to no collection or delivery on Sundays and during 12 holidays per year.

Note that the recovery objectives listed in Table 7.4 include recovered magnetic metals, iron and steel, which are divided into two separate products—light and heavy. The light includes mostly food and beverage cans and the heavy includes miscellaneous pieces of scrap. The reason for this separation is to meet two different markets [15].

The amount of recovered products expected, the last column of Table 7.4, can be used to construct the revenue statement, Table 7.5. This statement is divided into two columns, reflecting exchange prices and floor prices. The exchange price is the market price at which the recovered product is sold on a day-to-day basis. It may be the same as, or different from, the floor price—the guaranteed minimum sales price which may be negotiated for the product.

The values of the exchange and floor prices listed in Table 7.4 are based on actual LoIs to bid secured for a community [1, 2]. The exchange prices have been altered to reflect the decrease in the price of iron and steel scrap and an increase in the value of aluminum scrap since 1974, when the commitments were obtained. The sums of Table 7.5 can then be entered into the general cost and revenue statement for the indifference model, Table 7.6.

In Table 7.5, the price used for magnetic metals is in long tons (2240 lb), following industry practice. No floor price is shown for heavy magnetic metals or mixed nonferrous metals. Floor prices for these products were not obtained in the instance cited [2]. However, these products are always worth something to a scrap dealer.

In the case cited [2], advance market commitments were also obtained for handpicked corrugated and bundled newspapers, as well as recovered glass. In the general case of the indifference model used here, these products are omitted.

Table 7.6 is the generalized statement of revenues and cost of the proposed recovery plant. The capital cost is assumed to be $20,000 per daily U.S. ton of capacity, at the high side of actual experience [10, 12]. (The sensitivity of the indifference model to the capital cost is examined later.) The total capital cost of $20 million is assumed to be paid by 100 percent debt financing by general obligation bonds at $5\frac{1}{2}$

TABLE 7.5. Likely Revenues for Recovered Materials for Postulated
Recovery Plant (U. S. tons)

Material	Amt. available (tons per year)	Exchange price	Floor price	Annual revenue	
				Exchange	Floor
Iron and steel can stock	17,100 s.t. (15,268 l.t.)	$45/l.t.	$ 25	$687,060	$381,700
Heavy iron and steel	2,700 s.t. (2,411 l.t.)	55/l.t.	—	132,605	—
Aluminum cans and foil	1,050 s.t.	560/ton	160	588,000	168,000
Mixed nonfer- rous metals	300 s.t.	400/ton	—	120,000	—
		Totals:		$1,527,665	$549,700
				$5.09/input ton	$1.83/input ton

percent interest with even payback. The interest and payback are
shown for an equal annual payment retirement schedule. Operating
costs are assumed to be $10 per U. S. ton of input waste, based on
published estimates [11]. These estimates show that the operating
cost of a plant, on the basis of dollars per input ton, is nearly constant
for plants with capacities of from 500 to about 1200 U. S. tons per day.
In spite of this conclusion, the sensitivity of the indifference model to
the operating cost is examined later.

Table 7.6 contains entries for the revenue from the tip fee. The
essential feature of the indifference model is that the tip fee entered
for resource recovery is equal to the actual landfill disposal cost.
Although the current disposal cost may be entered, it is probably not
a proper comparison. The current cost may rise due to inflation
and/or imposition of more stringent environmental regulations govern-
ing the operation and use of the disposal facility. For illustrative
purposes, a value of $5 per U. S. ton is used as the indifference tip
fee in Table 7.6.

TABLE 7.6.   Generalized Statement of Revenues and Costs for the Postulated Recovery Plant (U. S. tons)

Capital Cost	Annual Values	
$20,000,000, even annual payments	$1,604,800	
Operating Costs		
$10/input ton	3,000,000	
Residue disposal, 73,500 tpy × $5/t	367,500	
Subtotal:	$4,972,300	
Revenues	**Exchange Price**	**Floor Price**
Materials Recovery	$1,527,665	$ 549,700
Tip fee, 300,000 tpy × $5/t	1,500,000	1,500,000
Sale of RDF (12 × 10^6 Btu/t)	?[a]	?[b]

[a] Plant shortfall (exchange price) = $4,972,300 − ($1,527,665 + 1,500,000) = $1,944,635. Shortfall is covered if RDF is sold for $0.90/$10^6$ Btu.

[b] Plant shortfall (floor price) = $4,972,300 − ($549,700 + 1,500,000) = $2,922,600. Shortfall is covered if RDF is sold for $1.35/$10^6$ Btu.

Table 7.6 does not include a revenue from the sale of RDF since it is difficult to obtain an advance LoI from an RDF customer. Also, unlike metals and paper, there is no established market for RDF, so it is difficult to price. (These points are established in Chapter 8.) Here is the advantage of the indifference model. Table 7.6 is now used to compute what price must be obtained for the RDF to maintain the indifference value of $5 per ton tip fee for both recovery and disposal.

Table 7.6 shows that if the tip fee is kept at $5 per ton, the recovery plant would have a financial loss. The RDF has to be sold for a price to at least cover that loss. This may be computed as follows (U.S. tons):

$$180,000 \text{ tpy} \times (12 \times 10^6 \text{ Btu/t}) \times \$?/(10^6 \text{ Btu}) = \$1,944,635/y$$

Solving this simple algebraic problem results in a price of $0.90/
(10^6BTU) ($0.009/MJ). This now is the <u>indifference value of the fuel</u>.
If the community can sell the fuel for this amount, resource recovery
costs the same as landfill ($5/ton). If the fuel can be sold for more,
obviously recovery costs less.

The indifference value of the fuel may be viewed as the point
where a community opens its negotiations with a potential customer to
secure a market and long-term commitment for the sale of fuel. The
first step is to determine what the potential customer is paying for
competitive fuel. There is little point in opening negotiations at the
indifference value if the customer is already paying less.

Several assumptions were made in developing the indifference
model. First, the capital cost of the plant may not be $20 million in
every community. The second is the estimate of operating cost.
These two assumptions are examined in Fig. 7.1 as a plot of the
indifference value of the RDF as a function of the capital cost of the
plant. A number of lines are shown, each representing a different
operating cost. Thus, for any capital cost over the range shown, and
for different operating costs, the indifference value of the fuel can be
determined. Note that these curves are computed for plants with materials revenue equal to the revenue from floor prices shown in Table
7.5.

Figure 7.2 examines the assumption of the amount of materials
revenue. The revenue is plotted as $/input U.S. ton, without differ-
entiating contributions from the various materials. Again, several
lines are shown, each for a different operating cost. The exchange
and floor prices from Table 7.5 are marked, showing that in the example of the postulated plant of Table 7.6, if everything "went wrong,"
the floor price would require a new RDF indifference value of $1.32
(10^6 Btu) ($1.40/kJ).

The indifference model may be used for planning recovery plants
for energy forms other than RDF. For example, it is possible to use
the model to determine the indifference value of steam or electricity
and compare it to market rates.

The indifference model permits a determination early in the
planning process of whether an energy product can be produced at a
competitive price under the constraint that resource recovery should
cost no more than current (or projected) methods of solid waste disposal. It further allows planners to refine the analysis quickly as
more data are developed.

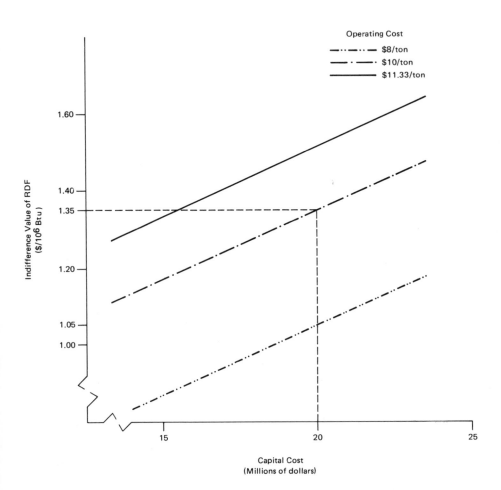

FIGURE 7.1. Relationship between operating costs and indifference
value of RDF (floor price assumption).

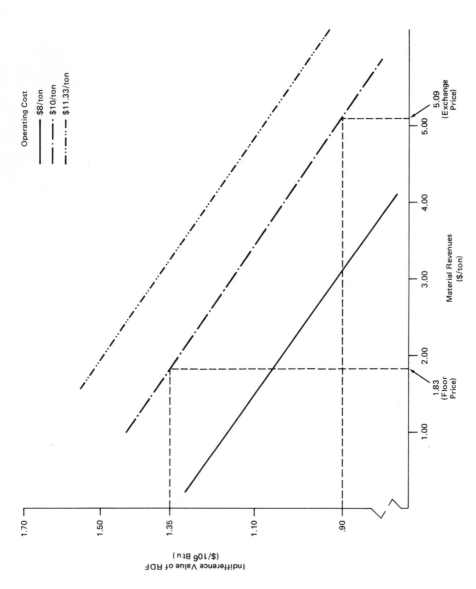

FIGURE 7.2. Relationship of material revenues and indifference value of RDF.

REPLACEMENT OF SCRAP VALUE OF
THE FACILITIES

The above indifference model has no sinking fund or other accounting provision for equipment replacement. Certainly, maintenance must include some equipment replacement due to wear and obsolescence. However, it is assumed that when the plant is depreciated (the bonds are retired), it will have no residual value.

By contrast, a landfill may have some residual value when it is retired. This value is limited to the extent that construction can take place. With present technology, this is a severe limitation. A landfill's usefulness may be further limited by its distance from population centers. Given these considerations, landfills, at least in the United States, have generally been thought of as future parks and other recreational facilities. It is difficult to place a real value on future parkland. Furthermore, funds will have to be spent to grade and plant the former landfill.

NOTES AND REFERENCES

1. H. W. Gershman. An approach to determining the economic feasibility of refuse-derived fuel and materials processing. In Proceedings, 1976 National Waste Processing Conference. ASME, New York, 1976, pp. 1-12.
2. National Center for Resource Recovery. Steps to resource recovery for the I-95 complex. Metropolitan Washington Waste Management Agency, Washington, D. C. , 1976.
3. There are perhaps six "laws of resource recovery." For a discussion, see H. Alter. Pitfalls when planning resource recovery. Waste Age 7(3): 12-14 (1976).
4. Environmental Protection Agency. Resource recovery and waste reduction, fourth report to Congress. Report SW-600. Washington, D. C. , 1977, p. 18.
5. There are several sources of information for estimating the quantity and composition of commercial and industrial waste. One which includes hospital and institutional dining facilities is U.S. Army Construction Engineering Research Laboratory. Pollution estimating factors. Technical Report N-12. Urbana, Ill. , 1976.
6. National Center for Resource Recovery. Materials recovery system: Engineering feasibility study. Washington, D. C. , 1972; National Center for Resource Recovery. New Orleans resource recovery implementation study. Washington, D. C. , 1977.

7. M. Biddulph. Principles of recycling processes. Conservation
    and Recycling 1: 31-54 (1976).
8. H. Alter and E. Horowitz, eds. Resource recovery and utiliza-
    tion. ASTM, Philadelphia, 1975.
9. National Center for Resource Recovery. Bid specifications for
    facilities and operation of the resource recovery and disposal
    program in New Orleans, Louisiana. New Orleans, 1973.
10. H. Alter. Energy recovery from solid waste: Looking through a
    dark furnace slowly. In Present status and research needs in
    energy recovery (R. A. Matula, ed.). ASME, New York, 1977,
    pp. 9-22.
11. R. L. Schroeder and B. M. Fabus. Resource recovery plant
    cost. In Proceedings, First World Recycling Congress (M.
    Henstock, ed.). Exhibitions for Industry, Oxted, Surrey, 1978,
    vol. 1.
12. P. J. Cambourelis. Resource recovery for municipal solid
    waste disposal. In Proceedings, Sixth Mineral Waste Utilization
    Symposium (E. Aleshin, ed.). IIT Res. Inst. and Bureau of
    Mines, Chicago, 1978.
13. J. G. Abert, H. Alter, and J. F. Bernheisel. The economics of
    resource recovery. Science 183: 1052-8 (1974).
14. R. E. Randol. Resource recovery plant implementation: Guide
    for municipal officials, financing. Report SW 157.4. Environ-
    mental Protection Agency, Washington, D.C., 1975.
15. H. Alter and W. R. Reeves. Specifications for materials re-
    covered from municipal refuse. Report EPA-670/2-75-034.
    Environmental Protection Agency, Cincinnati, 1975; H. Alter.
    Development of specifications for recycled products. Conserva-
    tion and Recycling 2: 71 (1978).

Chapter 8

MARKETING RECOVERED PRODUCTS

Little is to be gained from building a recovery facility if the recovered
materials or energy product (e.g., fuel) cannot be sold and/or used.
Before investing in a facility, a municipality (or its agent) should have
reasonable assurance that markets for the recovered products exist.
Markets come first and specifications determine technology. Thus,
securing assured markets should be the first step in planning a facility. This is easier for recovered materials than fuel or other energy
products. Recovered materials resemble and can substitute for established grades of secondary or virgin raw materials. The same is not
always true for the energy product. Steam and electricity are similar
to products produced by fossil fuels. However, steam and electricity
must also meet specifications. If the product is refuse-derived fuel
(RDF), there are few guidelines and little experience on which to base
the marketing, sales, specifications, or prices.

Municipalities generally are accustomed to buying, not selling.
Procedures and regulations for disposing of surplus property, such as
old vehicles and excess land, are likely to be cumbersome for selling
"surplus" materials recovered from wastes.

This chapter is intended as a guide for locating and securing markets for recovered materials and energy products. The emphasis is
on the former simply because there is more experience and buyers
have been more willing to give long, binding, advance commitments.
These commitments have often been for the first 5 operating years
of a facility, which often translates into 9 or 10 calendar years due to
the time required for planning, design, and construction. Some municipalities have sought sales agreements for the amoritization period
of the plant, perhaps 20 or more years. Furthermore, they have
sought such commitments on a take-or-pay basis, especially for the
energy product. Such arrangements would facilitate financing, but
they are unrealistic. The municipal manager merely has to put him-
self in the role of the business man to appreciate this.

111

The desire for long-term, take-or-pay contracts illustrates an interesting element of municipal practices and a difficulty of resource recovery planning. Municipalities seek a risk-free contractual arrangement. However, resource recovery is a production enterprise—the production of materials and fuels to specification—and subject to normal business risks.

MARKETS FIRST

The cost of recovering materials from municipal solid waste can be recovered by revenue from two sources: sale of recovered by-products and tipping fees.* The latter is often a user service charge or a line item in the municipal budget. Management prudence and public demand seek to keep the tipping fee minimal and unchanged over several years of operation. The fee can be approximated over the first few years of facility operation only to the extent that the sales revenue from recovered products can be estimated for the same period. Therefore, a facility's financial forecasts are based on market commitments for recovered products used in conjunction with capital and operating costs. The required tipping fee is simply the amount necessary to ensure economic viability.

Locating potential customers and concluding several years of market commitments in advance of plant design have been achieved by several cities. This marketing may be performed by a municipality, its consultant, or a potential contractor. Two basic approaches have evolved: the first relies on systems suppliers to secure their own market as part of their bid to build, own, and/or operate a proposed facility. The second attempts to secure these commitments in advance of a system procurement. In the latter case, the marketing is performed by the public sector or its consultants. The markets are then assigned to the systems winner if he is to operate the facility, or used by the public sector if the facility is operated by a governmental entity.

The following approach is directed at obtaining binding letters of intent (LoIs) that eliminate or greatly reduce the risk of selling recovered products and establish realistic sales forecasts prior to budget authorizations.

---

*Also called dump or disposal fees.

MARKETING AND SALES STRATEGIES

The poor reputation of materials recovered from waste stems from past performances. Sufficient quantities of materials of the required quality often were not produced and delivered on schedule. The operator of a resource recovery plant, therefore, must make certain assurances prior to marketing. Potential customers have to be assured of product quality and delivery as for other raw materials. The burden is clearly on the prospective producer.

The quality of the recovered product (its specification) must assure utility in current manufacturing operations. At the outset, industries that use certain raw materials cannot be expected to substantially alter their processes to accommodate recovered material. A resource recovery plant must establish quality control, testing, and measurement programs. Failure to meet specifications will reduce product quality and can result in lower prices and rejection of shipments. Finally, sellers should guarantee the quantity delivered over a period of time, since contractual delivery schedules are common and necessary. Customers must receive raw material on schedule to continue production.

These commitments may seem restrictive, but they are common in the raw material supply industry. Resource recovery may seem foreign to municipalities more accustomed to service functions. If, therefore, a community is unprepared to offer product assurances, it should consider delegating marketing and production to the private sector. There are several sales strategies for marketing recovered materials.

Don't try to sell garbage. Products must be clean and to specification. Recovery itself is not a basis for sale, and recovery processes will have to meet specifications. Testing should be employed to determine whether recovered materials meet specifications. It would be well to decide which tests are to be used early in the marketing process.

Take advantage of current processes. Markets for recovered material are more easily established when the product is employed in an existing process than when used in a process designed specifically for the recovered product. Also, the product's demand will be greater if it is similar to an existing secondary material.

Meet local markets. There are two reasons for this. First, it reduces freight costs; second, a seller must offer customer services. Contracts between buyer and seller are best achieved when transportation and communications distances are minimal. Thus, long-distance calls, personal travel, and lost freight shipments are reduced. These factors are worthy of consideration in discounting gross selling prices when compared to distant markets.

Don't overoptimize recovery processes. Due to public pressure, there is a natural tendency to maximize revenues rather than return on investment or other measures of investment success. Maximization of revenues means recovery followed by upgrading or conversion to a higher gross value product. It seems unwise to add processing steps if they can be done by a scrap processor or other existing industry.

Beware of price optimism. There are several published lists of prices paid for secondary metal and paper materials. They refer to standard grades of secondary materials, not necessarily the products recovered in a resource recovery plant. For example, the price of can scrap may be tied to no. 2 scrap steel bundles, but old cans are clearly not no. 2 bundles and may be worth less. Aluminum recovered from municipal solid waste is not equal to the standard scrap grade of "smelters' old sheet." Furthermore, the quoted prices are for the "spot" market. Clearly, a quid pro quo for a long-term commitment is an exchange price at some discount from the spot prices. To "play" the spot market is to risk having large inventories.

Beware of market penetration costs. Some planners, perhaps seeking to maximize gross revenues, have considered adding product manufacturing facilities to the resource recovery plant, such as a wallboard plant utilizing recovered paper, a cinderblock plant utilizing recovered glass, an ammonia or methanol plant, and so on. Added costs include a sales force, distribution system, and advertising budget to sell what may be untried or unknown products. Also, consider the effect when the competitor's salesman tells a customer: "It's made from garbage." Overcoming the new product's lack of acceptance is a market penetration cost which may be high.

THE MARKET SURVEY

In planning resource recovery facilities, two types of market surveys are often conducted. The most common and uncritical is to locate users of steel, paper, glass, and so forth within a certain radius. The second is to start the approach recommended here. It is based on familiarity with local industry, the ability of known resource recovery processes to produce materials to simple specifications, as well as employing the marketing and sales strategies discussed above. There are likely local markets for each recoverable material. Matching these to local conditions, combined with negotiation of advance commitments, constitutes the market survey.

The limited local survey is followed by obtaining LoIs for the purchase of recoverable materials. They are intended to obtain an assured financial forecast for the proposed facility. One LoI per material can be sufficient, and it is the key element in this approach.

Prospective markets for recoverable materials from MSW are listed in Table 8.1. It offers a starting point for identifying which local businesses to contact. The trade-offs of selling products to dealers and scrap processors, as opposed to primary producers, will be discussed separately.

After locating prospective markets, the next step is to approach the potential buyers, remembering the plant is not built and there are no samples to show. A strong and binding advance commitment to purchase is sought. At this point, a specification—not a product—is being sold. The specification is a designation of particulars of product form and composition to be agreed upon by buyer and seller.

Chances are the buyer is unfamiliar with specifications for products recovered from refuse but has standards for virgin and secondary materials. The seller should familiarize the buyer with available specifications for recovered materials as a starting point for negotiations. Specifications for recovered materials are discussed in Chapter 9.

ROLE OF THE SCRAP PROCESSOR

The metal scrap processor, or paper dealer, generally buys to a lower specification and at a lower price than the eventual user of the secondary product. In return, he performs two valuable services. One is the ability to spread the recovered material into several markets and/or to several customers within a market. This helps to balance the economic demands and price changes over a long period. The second service is upgrading low-value scrap—that is, processing mixed paper or dirty cans—to grades suitable for sale.

In return for these services, the scrap processor receives a fee which lowers the gross price received by the resource recovery plant compared with what may be paid by paper and steel mills or other primary customers. However, the reduced purchase price must be evaluated by the municipal planner in light of the value of the services and the consequent reduced investment in resource recovery facilities and operation.

Until recently, with few exceptions, scrap processors and dealers were unable to offer prospective recovery operators advance commit-

TABLE 8.1. Prospective Markets for Recovered Materials

Material	Market and Use	Remarks
Newspapers	Deinking for new news; as-is for chipboard	Most old news sold through dealers. Some can be sold directly to deinking plants.
Separated corrugated	New corrugated, kraft paper, or linerboard	To dealers or mill. Few mills in urban areas, hence function of dealer.
Old cans	Detinning Blast furnace (new iron) } sell direct or to Steel mills } scrap processor Foundry (not yet commonplace)	Few detinning plants. New ones can be built locally if quantity is sufficient.
	Ferroalloy Copper mines (best through dealer)	Ferroalloy industry only in a few areas.
Other steel scrap	Scrap processor	Material a complex mixture of alloys and types. Need a scrap processor.
Aluminum cans	New can stock—to primary producers Secondary users—to primary producers and dealers, including secondary smelters	Sold F.O.B. recovery plant.
Miscellaneous aluminum	Secondary uses	Salable to dealers and secondary smelters.
Mixed other nonferrous metals	Scrap processors	Mixture must be processed by specialists in the scrap industry.
Glass cullet	Glass container manufacturers	Must be ultraclean. Broader market if color-sorted.
Glas sands	Brickyards—for use as a flux	Market not yet established. Great interest shown based on preliminary experience.

ments to purchase the recovered materials. Scrap processors often
do not have the financial resources of the primary materials producers, hence they cannot offer the same type of binding LoI.

THE LETTER OF INTENT (LoI)

The LoI is the instrument negotiated between the resource recovery
planner and the purchaser of recovered materials. The culmination
of the market survey, it financially underpins the resource recovery
plant and precedes the buyers' purchase orders.

Frequently, planners have been willing to accept weak and non-
commital expressions of interest by unsure, unconvinced, or uninter-
ested buyers. Such expressions are worthless since they cannot
assure that the plant's output will be sold. Moreover, there are cases
where properly binding commitments have been negotiated. Some
aspects which must be considered in negotiating such advance commitments are given below.

The fundamental terms and conditions in the LoI are <u>length of
commitment, quality, quantity of material, delivery schedule, termination,</u> and <u>price.</u> The LoI may be worded as an intent to issue a
purchase order to the resource recovery plant subject to the terms
and conditions set forth in the advance commitment.

The <u>length of commitment</u> should be long enough to allow the recovery plant operation to start and gain momentum. For most materials it is unreasonable to expect the advance commitment to cover the
full financing term, which is likely to be 10 or more years, unless
the operator is prepared to accept a discount on the selling price over
the term. Because of the time needed to plan and construct a recovery
plant, even a 5-year operating LoI may translate to 8 or 9 calender
years.

The LoI should specify the amount of material to be sold. The
quantity, of course, cannot exceed the estimated amount of the material in the waste, multiplied by the quantity of waste committed to the
resource recovery plant, multiplied by the estimated recovery efficiency. Because of the processes' uncertain combined productivity,
the quantity of delivered recovered material may be expressed as a
range for the first year, subject to adjustment within this range after
the first operating year.

The <u>quality</u> of the material to be delivered is delineated by a
specification which becomes part of the LoI. Furthermore, the LoI
should address whom will be responsible for the accompanying test
and quality control program and the bases for rejection or downgrading.

Operators of facilities must guarantee <u>delivery</u> of a specified quantity. The operator may not divert deliveries to "spot" markets, even at higher prices, unless allowed in the LoI. The delivery schedule is based on quantity per day, week, or month. The form of transportation and minimum shipments are also stated. This is essential information for planning the storage facilities, shipping docks, railroad siding, or truck unloading facilities at the users' plants.

Termination clauses must be fair. Obviously, the seller cannot expect the customer to continue if the plant closes. The buyer cannot reasonably terminate if the product is merely suspected of interfering with some operation or quality point. <u>Force majeure</u> should be given its usual narrow interpretation. There is yet another aspect of termination. The contract price for scrap agreed upon in the LoI may occasionally exceed the spot market price. There is a natural temptation for the seller to take advantage of such situations, but this will deny the buyer assured quantities of material. A fair compromise is for the LoI to permit the seller to annually seek a higher price during years 2 through 5. The original buyer would have the right to match the higher price. If the seller deals with another buyer, it must be for a reasonable length of time, perhaps a minimum of one year. In return for guarding the original buyer against the seller playing the spot market and disrupting deliveries, the original buyer agrees to take back the seller, at the original terms, should there be an absence of one or more years with another buyer. Obviously, this is a give-and-take arrangement that permits the seller to seek a higher price while protecting the original buyer in return for his other guarantees.

Because salable recovered material will substitute for, or be in addition to, standard scrap grades, the LoI means that the recovery plant is first in line to purchase as the buyer's scrap needs may be diminished—not last, as is often the case with new sources of supply.

Pricing Arrangements

There are several ways the exchange price—the price paid for the recovered material—can be established and expressed in the LoI. If the price is not fixed in the LoI, a floor price should be obtained in addition to the exchange price relationship. The floor price is an essential element in the financial forecast that justifies investment. Alternatively, prices for some recovered products (steel and paper in particular) may be tied to scrap quotes for grades and sections of the country as published in standard periodicals.

A third pricing arrangement pegs the buyer's price to sales for like materials from other sources. This is useful when the buyer has no access to historical data and the recovered product is a small amount of total raw material purchased. This arrangement has been used for newsprint when the purchasing mill and its largest supplier were not covered by quoting services. Presumably, it is the basis for determining market prices for glass cullet when prices are determined by published costs of equivalent raw materials (usually sand and soda ash).

A classic bargaining question is whether to trade some of the "upside" benefits for a lessening of the potential "downside" risks. For example, the seller may have to choose between accepting a price of no. 2 bundles but a $10-per-ton floor price—or a fraction of the no. 2 bundle price (say 65 percent) with a $20-per-ton floor price.* The choice is judgmental but highly influenced by the need to minimize risk. In the latter pricing arrangement, the buyer is entitled to a 35 percent (or other) discount in exchange for offering a reasonably high guaranteed floor price. During periods of high prices, the buyer benefits from lower priced materials; during times of low prices, the seller benefits by having a guaranteed buyer.

These suggested pricing mechanisms are subject to negotiation and may provide an incentive discount when negotiations bog down. (An incentive discount might be: "Sign now and I'll drop the price $1 per ton.") Some of the periodicals which regularly publish scrap quotations are listed in Table 8.2.

Public sector decision makers may be more attracted to floor prices than exchange prices when considering implementation of a resource recovery facility. The floor price, even if lower than the market or exchange price, offers a no-risk opportunity. Thus, the planner or consultant arranging for LoIs should perhaps negotiate higher floor prices and lower exchange prices. A similar situation may apply to a private sector representative negotiating the LoIs on behalf of a public sector client.

PRIVATE VERSUS PUBLIC OWNERSHIP

The above discussion of terms, conditions, and pricing is generally applicable to both private and publicly owned resource recovery facilities. However, private owners have the right to negotiate their arrangements and prices independent of considerations other than the final price bid or charged to the municipality.

---

*Prices for steel scrap are always quoted in the United States for a long ton, 2240 lbs.

TABLE 8.2. Published Scrap Quotations

American Metals Market (ferrous and nonferrous metals)
7 East 12th Street
New York, NY 10003

Iron Age (ferrous and nonferrous metals)
Chilton Company
Radnor, PA 19089

Official Board Markets (paper and paperboard products)
Magazines for Industry
20 North Wacker Drive
Chicago, IL 60606

Secondary Raw Materials (all secondary materials)
Market News Publishing Corporation
156 Fifth Avenue
New York, NY 10010

If the plant is publicly owned, the dicta of fairness and open government generally require that all responsive and responsible parties can bid for the recovered material. The challenge, then, is to secure the LoI in advance while preserving public bidding rights. An innovative solution is a letter of intent to bid (as distinguished from a letter of intent to buy). A sample LoI to bid is provided in Appendix C.

By signing a LoI to bid, the purchaser agrees to submit a response to an invitation to bid for the purchase of recovered material at some time in the future. The LoI to bid covers all of the necessary terms, conditions, and price structures. The bidder further agrees that the offer will not be less than a stated price. Generally, the floor price is specified in the request and the exchange price (or exchange price relationship) is the competitive aspect of the response. These prices can then be used for financial planning, just as an LoI to buy. Of course, prices will be lower in an LoI to bid since it is only the first step in product procurement. The bidder may increase the price in a final invitation to bid when actual rights to the product are at stake.

There are two cautions in dealing with an LoI to bid. First, keep exact prices in the LoIs confidential prior to final bid opening. Knowledge of a firm's price could enhance the position of a competitor.

This may be arranged confidentially through a consultant or other third party. Second, the legality of this approach has not been tested. However, it seems reasonable, fair, open, and within the spirit of public bid laws.

CANCELING THE LoI

Municipal planning for resource recovery is a lengthy process, often taking more than three years. It is unfair for buyers to maintain a commitment for this period unless there is a reasonable chance the plan will be implemented. A potential buyer's commitment is generally based on obtaining specific amounts of material from a single source. Therefore, the LoI (whether to bid or buy) should include a statement terminating the commitment unless the municipality has demonstrated substantial progress by a certain date. Substantial progress may mean completion of a planning document, issuance of a request for proposals, or similar event. The commitment may be nonexclusive and based on which of several communities completes its arrangements first.

IMPLEMENTING THE MARKETING

Obtaining advance commitments for the purchase of recovered materials is the most important step in resource recovery planning. Commitments provide some of the advance financial assurances municipal managers seek, and the accompanying specifications determine major aspects of the proposed plan. Clearly advance commitments are not ironclad. There are some market risks as with any production enterprise.

The steps leading to a set of LoIs (to bid or buy) should be undertaken by technically knowledgeable and skillful negotiators. It is a time-consuming process that often exceeds the capacities of competent municipal managers. Marketing will probably have to be delegated to someone else, which may be done in several ways. First, the municipality may engage a consultant to conduct marketing and provide LoIs to bid. The community then completes the bid process and the price relationship and specifications of the winners are integrated into the planning and implementation process. This can be done in both public and private ownership and operation.

Second, potential bidders may provide LoIs in their bids. The municipality should delineate minimum terms and conditions and possibly provide the form of the LoI to ensure that the bids are comparable. This arrangement more readily allows use of proprietary recovery processes than the former.

Finally, under plans for a privately owned resource recovery facility, the municipality may merely require bidders to provide an LoI to buy—according to minimum terms and conditions—as part of the evidence of ability to perform. An alternative is self-insuring the market for some products, a step which can be done by some large conglomerate companies.

Passing the burden of market attainment to the private sector has fundamental appeal. The municipality need only be concerned with the result of a well-constructed LoI and the net tipping fee, and can ignore many in-between steps. The private operator can be obligated to meet product specifications. Thus, he must produce a salable product, and the municipality is largely relieved of operational risks. However, some drawbacks are apparent. This approach goes hand-in-hand with issuing a serious request to bid and operate a facility and cannot be used to test the market.

Preparing a satisfactory request for proposals is an expensive undertaking. Considerable time and resources are required by all parties involved. Likewise, from the private sector's standpoint, preparing the response involves much more than the marketing effort. Therefore, unless there is a rather strong level of commitment from the public side, it is unlikely that respondents will make a maximum effort to prepare proposals and obtain market assurance.

APPLICABILITY TO ENERGY PRODUCTS

The general rules of marketing recovered products are more broadly applicable to recovered materials than to energy products because recovered materials are similar to standard scrap grades. Also, operators of resource recovery plants are more anxious to obtain long commitments for sale of the energy product, perhaps as long as the length of the amortization period of the plant, to help secure financing and to avoid alternative disposal methods. The energy product is the result of processing and disposal of the organic portion, by far the largest increment in terms of weight and volume.

Generally, the experience in obtaining LoIs for energy products has not been as good as for materials. Following are some additional requirements for securing such markets.

STEAM AND ELECTRICITY

If the recovery facility is to generate and sell steam and/or electricity, these products must also meet the users' specifications.

Specifications for steam include temperature, pressure, saturation, and availability. Specifications for electricity include voltage, phasing, and availability. Buyers of electric power often insist on providing the design of the interfacing switchgear. Buyers of both steam and electricity may have specific delivery requirements when waste is unavailable or the recovery facility is inoperative. In short, the operator may have to install standby capacity, such as an oil-fired boiler.

Users of steam and electric power have periods of peak demand and baseloading requirements. Peak demand might be at certain times of the day; for example more electric power is consumed in mornings and evenings than late at night. More steam might be needed in the cold winter and hottest parts of the summer (for air conditioning) than in the spring and fall. If the recovery plant is a major source of steam or power, it must manage its production to meet such schedules. Furthermore, peak power is worth more than baseloading power. Some European heat recovery incinerators generate only a portion of their power from waste. In this way, the fossil fuel takes up the swings in demand over the year.

The difference between interruptible and noninterruptible customers is another feature of selling steam and electricity. An example of the latter might be supplying steam for district heating and air conditioning. Under such circumstances, the steam must be available 7 days a week, 24 hours a day. Steam to this customer is worth more than to the interruptible customer, such as an industrial plant, which might purchase only a portion of its steam (or power) needs from the recovery plant, generate the remainder itself, or purchase it from a utility.

An excellent example of the latter case is the waste-to-steam plant in Akron, Ohio, scheduled to be commissioned in 1980. The municipal district heating system, including a local hospital and university, are noninterruptible customers. A local tire-manufacturing plant purchases excess steam at a lower price, uses it for baseloading, and generates additional steam with oil. In this way, the waste-to-energy plant sells all its steam.

In planning to sell steam, it is important to gauge the variations in amount of refuse (fuel) available over a year and compare the result to the demand over the year. There must be some sort of match. Additional steam could be generated from conventional fuel, but excess steam will only be condensed, or excess waste will be landfilled.

Some communities in the United States and Europe have built heat recovery incinerators to generate steam with the hope of establishing an industrial park or other steam customer on an adjacent site. Few of these have been successful. However, they illustrate a frustrating "chicken-or-egg" situation. The industrial park users cannot build and start operations until steam is available. The steam plant should not be built until the customers are assured. Some form of contingency planning, where both parties agree to build only in conjunction, might resolve the problem. Surely, this requirement imposes an additional degree of complexity, but it eliminates the expensive embarrassment of a steam-generating plant with a large condenser and no customers.

The sale of electricity is somewhat simpler. The amount of electric power likely to be generated by a waste-to-energy plant is small compared to the amount a utility generates and/or could buy from the grid. Nevertheless, the utility will likely have requirements regarding peak and baseloading due to commitments to other suppliers. Furthermore, federal or local utility laws may require that the generating equipment be owned by a utility instead of the waste-to-energy plant.

REFUSE-DERIVED FUELS

The sale of refuse-derived fuels, such as fluff or densified RDF, to private bodies such as manufacturing plants, should be as straightforward as selling recovered materials. Such potential customers usually are seeking a usable fuel at a low price. For example, a cement manufacturer in England obtains waste from the local authority and processes it to RDF by a technology similar to scheme (1) in Chapter 5. The RDF is a lower-cost alternative fuel for the cement kilns. Thus far, there are few such examples. Most RDF sales have been to utilities having some unusual requirements.

Utilities have an obligation, if not legal requirement, to furnish power to customers reliably and at the lowest possible cost. Any feeling that RDF will compromise these requirements impedes the letter of intent's ratification and often results in special clauses.

The special clauses often attempt to minimize the customer's risk. These have taken three forms: (1) the net cost of using RDF for the generation of electricity should be no more than if coal was used; (2) if RDF use causes technical or economic difficulty, the utility can unilaterally terminate the purchase agreement; and (3) a difficult-to-meet specification.

The problem of difficult specifications appears to be diminishing. In the past, some utilities imposed restrictions in the form of specifications of properties which cannot be controlled. For example, one utility specified the nitrogen content of the RDF. The nitrogen content of RDF in the United States is less than 1 percent by weight, according to the examples in Table 5.3. Furthermore, there is no known technology for reducing or controlling it. Thus, by specifying the nitrogen content, the buyer created an excuse for rejection. Fortunately, there appear to be few examples of such specifications.

In many RDF sales agreements, the customer pays the going rate for the fuel on a calorific or heat value basis, less incremental costs. These incremental costs may include repayment of the capital cost of RDF storage and handling facilities, modifications to the boilers and air pollution control equipment, increased expenditures for operation and maintenance, and economic dispatch penalties. While it may seem a fair arrangement, it has some drawbacks for the seller. First, it is unclear if similar arrangements are made when the utility switches to any other fuel. Second, only the buyer fully understands and controls these factors. The seller cannot easily enter into cost and pricing negotiations on an equal footing with the experts. Even though these factors are regulated and open to public inspection, they can be difficult for people outside the industry to grasp.

The economic dispatch penalty is a case in point. Economic dispatch basically refers to the practice of first using the lowest-cost generating units. During periods of baseloading, the least expensive to operate units will be on-line. As more power is needed, the next most expensive unit is turned on, and so forth. In this way, the utility is able to minimize the average cost of electricity. In some systems, however, a nuclear powered facility may be the least expensive and the older coal-burning units the most expensive. RDF will likely be used in the older units. Therefore, the utility will seek a discount in the selling price of the fuel to equalize the cost of power generation in the various units. This is an economic dispatch penalty to the RDF seller which can be considerable. There is at least one case of a utility telling a municipality that it will require to be paid to burn the RDF in order to even out dispatch cost between old coal-burning units and the new baseloading nuclear units.

Few would argue the fairness of economic dispatch, certainly not utility customers who seek the minimum average cost of power. However, it is a complex adjustment or selling price discount to negotiate and audit. (An alternative means of pricing will be discussed later.)

Economic dispatch can also be the basis for acceptance, suspension, and termination of the fuel's use. For example, one utility agreed to purchase and use RDF provided that the generating station operated at 50 percent or more of design capacity. This economic dispatch is again under the sole control of the buyer.

Utilities have argued for unilateral suspension privileges as soon as they believe the RDF harms their units. Such privilege would disrupt the operation and finances of the waste-to-energy plant. The sellers have also sought most favorable treatment. For example, some have insisted on take-or-pay contracts for the RDF. Such con-

tracts would help insure the financial success of the plants and provide
an alternative to disposal. However, they could also disrupt the
buyer's business. In case of economic or technical difficulties with
RDF, or even normal shutdowns for maintenance, the buyer would be
paying for unusable fuel which has to be disposed of.

AN LoI FOR RDF BUYERS

The LoI approach to sale of recovered products should be generally
applicable to marketing RDF or other energy forms. Termination can
be dealt with by suspending deliveries and RDF use. If it appears a
RDF is harmful to boilers, the user should have the right to unilaterally stop delivery and use. This suspension would continue until the
cause is ascertained. The LoI and succeeding contract should have
provision for the use of independent consultants or a similar arbitration method to resolve differences. If it is agreed that continued use
of RDF would harm the boilers, the user should have the right to
terminate.

There are, perhaps, two fair and simple ways of establishing
selling price. One is to establish a fuel value based on calorific value
or heat content, adjusted annually according to the buyer's costs as
reported to the regulatory bodies. This might be done, for example,
by using the cost of electric generation at one station as a base for
establishing the cost of generation at another station using RDF. The
base can be determined from historical records. For example, the
past ratio of generation costs at two stations can be used as the adjustment factor for determining the price of RDF. The calculation could
be made annually, quarterly, or for other intervals. Thus, all costing
elements are combined into a single index.

A second, simpler method would be based on the indifference
value of the fuel (determined according to the model in Chapter 7)
compared to the average price the buyer is paying for coal. If the
indifference value is less than the market price for coal, a simple
compromise may be to split the difference. This method is not analytical, but it may be effective in reaching an agreement. Of course,
the two methods can be combined.

UPGRADING THE FUEL

Reputations of higher moisture and ash contents have caused users not
to consider RDF. It is possible to process RDF to lower both and
improve its marketability or acceptability.

The ash content may be reduced using processing schemes such as (5), (6), or (7) in Chapter 5. These include a screening step—either before or after shredding—to remove glass, stones, and similar inorganic fines that contribute to the ash fraction. Doing so decreases the mass of the product, increases the amount of waste material, and reduces the amount of ash the buyer must dispose of. Furthermore, removal of ash components raises the calorific value of the fuel delivered. Ash is inert, contributes nothing, and detracts a little from the usable calorific value in that it is heated during combustion and its heat content is wasted when the ash is cooled prior to disposal. The buyer usually pays on the basis of calorific value received, so is likely to pay more per mass of fuel received after ash removal. The seller will realize more or less net revenue depending on the cost of disposing of the ash.

Screening some form of RDF prior to delivery removes only a portion of the ash. The cellulose and other products constituting RDF contain inherent ash which cannot be removed by screening. Examples are fillers in paper, pigments in coatings, and inorganic materials naturally contained in cellulose.

Removing moisture has a greater effect on the calorific value of the fuel than removing ash. Mositure, like ash, contributes nothing to the calorific value, but fuel energy is consumed by vaporizing it. Also, a "wet" fuel generates larger quantities of combustion gases which must be handled and processed to remove pollutants, at a higher cost than for "dry" fuels.

There is a question whether it pays to dry RDF prior to sale. Most builders and designers of plants have decided not to; others prefer to remove some moisture. RDF is a low-density, poor heat conductor, so drying such material is likely to be inefficient. The "cost" of drying is calculable in terms of the energy likely to be consumed expressed as a portion of the RDF produced.

The RDF has a heat content per unit weight of $J_c f$, where J_c is the calorific value or energy released on combustion and f is the fraction of RDF that is fuel and inert (ash). The moisture content is (1 - f). To remove this moisture (again, per unit weight) requires a heat input of

$$J_w(1 - f) + (1 - f)C_p(T - 373)$$

where J_w is the heat of vaporization of water from room temperature to the boiling point, C_p is the heat capacity, and T the final absolute temperature of the water vapor. Because $J_w \gg C_p(T - 373)$, the amount of heat added to dry may be approximated as $J_w(1 - f)$.

The cost of removing the water, in terms of the fraction of RDF used, x, can be calculated by balancing the heat added to dry against a portion of the fuel for any given heat exchanger:

$$J_c \cdot x \cdot E = J_w(1 - f)$$

where E is the efficiency of the drying (efficiency of energy utilization in vaporizing the water). Solving for x,

$$x = \frac{J_w(1 - f)}{J_c \cdot f \cdot E}$$

This expression is plotted in Figure 8.1 as the variation of x, the fraction of the fuel needed to dry the remainder, as a function of the moisture content $(1 - f)$, for different values of J_cE over a range of interest. Note that the fraction of fuel required, x, increases sharply as the moisture content increases. In short, drying RDF can consume a large fraction of its deliverable energy.

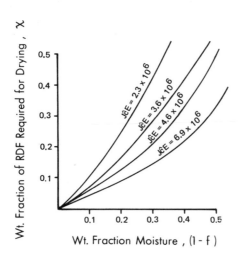

FIGURE 8.1. Relation between the amount of RDF needed to reduce the moisture content of the remainder, for different values of the RDF calorific value, J_c, and drying efficiency E. (Source: H. P. Sheng and H. Alter. Energy recovery from municipal solid waste and method of comparing refuse-derived fuel. Resource Recovery and Conservation 1: 85-93 (1975).)

BIBLIOGRAPHY

Alter, H. Development of specifications for recycled products.
    Conservation and Recycling 2: 71-84 (1978).

Alter, H. and E. Horowitz, eds. Resource recovery and utilization.
    ASTM, Philadelphia, 1975.

Alter, H., S. L. Natof, K. L. Woodruff, and R. D. Hagen. The
    recovery of magnetic metals from municipal solid waste. Report
    RM 77-1. National Center for Resource Recovery, Washington,
    D. C., 1977.

Alter, H. and W. R. Reeves. Specifications for materials recovered
    from municipal refuse. Report EPA-670/2-75-034. Environ-
    mental Protection Agency, Cincinnati, 1975.

Duckett, E. J. Contaminants of magnetic metals recovered from
    municipal solid waste. Report NSF/RA-770244. National Center
    for Resource Recovery, Washington, D. C., 1977.

Duckett, E. J. The influence of tin content on the reuse of magnetic
    metals recovered from municipal solid waste. Resource Recov-
    ery and Conservation 2: 301-28 (1977).

Garbe, Y. M., and S. J. Levy. Resource recovery plant implemen-
    tation: Guide for municipal officials: Markets. Report SW-157.
    Environmental Protection Agency, Washington, D. C., 1976.

Gordian Associates. Overcoming institutional barriers to solid waste
    utilization as an energy source. Report HCP/L-50172-02. U. S.
    Dept. of Energy, Washington, D. C., 1977.

Lowe, R. A., ed. Use of solid waste as a fuel by investor-owned
    electric utility companies. In Proceedings, EPA/Edison Elec-
    tric Institute Meeting. Report SW-6P. Environmental Protection
    Agency, Washington, D. C., 1975.

Randol, R. E. Resource recovery plant implementation: Guide for
    municipal officials: Risks and contracts. Report SW-157.
    Environmental Protection Agency, Washington, D. C., 1976.

Sheng, H. P., and H. Alter. Energy recovery from municipal solid
    waste and method of comparing refuse-derived fuel. Resource
    Recovery and Conservation 1: 85-93 (1975).

Suloway, M., and T. V. Sprenkel. Organization, financing, and
    management in solid waste disposal systems: Chicago, a case
    study. Paper presented at the Symposium on Solid Waste Con-
    version to Energy, Hamburg, 1977.

Chapter 9

SPECIFICATIONS FOR RECOVERED PRODUCTS

A specification for a recovered material or fuel is the designation of particulars of form and/or composition and is the basis for acceptance or rejection, for sale or use. The specifications for a raw or processed material may also be used to determine its selling price.

Specifications are used in most if not all forms of commerce. The production and sale of products recovered from mixed municipal wastes cannot be exceptions. Noteworthy, the development of specifications for recycled products is not new. The secondary materials industry in the United States, as an example, has employed specifications for approximately 65 years, since the founding of what is now known as the National Association of Recycling Industries. The Institute of Scrap Iron and Steel pioneered the establishment of specifications for scrap ferrous metals approximately 50 years ago. The specifications of these organizations now form the basis for the published commodity quotations of many grades of scrap or secondary materials. Table 9.1 lists the source of such specifications in the United States, and Table 8.2 lists sources of commodity quotes of secondary materials based on these.

As the concept of resource recovery from mixed municipal wastes grows in importance as an industrial activity, new and—as will be shown later—different forms of specifications must be developed to meet the needs of this new activity. Several general criteria apply to developing specifications for secondary materials. These are discussed below.

Note: This chapter is based largely on a recent discussion of progress in the field [5] .

TABLE 9.1. Sources of Specifications for Recovered Materials

All materials	Glass

ASTM Committee E-38,
    Resource Recovery
American Society for Testing
    and Materials
1916 Race Street
Philadelphia, PA 19103
	Glass Packaging Institute
	1800 K Street, N.W.
	Washington, DC 20006

	Nonferrous metals
Steel	National Association of Recycling Industries, Inc. 330 Madison Avenue New York, NY 10017
Institute of Scrap Iron and Steel, Inc. 1627 K Street, N.W. Washington, DC 20006	
	Aluminum
Paper	Aluminum Association 818 Connecticut Avenue, N.W. Washington, DC 20006
American Paper Institute 260 Madison Avenue New York, NY 10016	National Association of Recycling Industries, Inc. 330 Madison Avenue New York, NY 10017
Paper Stock Institute of America 330 Madison Avenue New York, NY 10017	Aluminum Recycling Association 900 17th Street N.W. Washington, DC 20006

The specifications should permit that the recovered materials be processed for sale in a form and degree of cleanliness that will require minimum perturbation to user industries [1]. The specification should permit forms easily produced, shipped, and handled from a variety of resource recovery processes. It should not depend on only one or a few types of processing, particularly proprietary methods. Further, the specification should be capable of being tested using existing (and hopefully simple) analytical techniques. Specifications must be within the state of the art of known recovery systems and processes and as quantitative as possible. Nonspecific terms such as "clean" should be avoided, as should absolute terms such as "resistant."

Some products likely to be recovered from mixed municipal solid waste (MSW) in industrialized countries are paper, steel, aluminum, glass, and a mixture of organic materials as some form of refuse-derived fuel. There have also been attempts to separate and recover plastics for reuse. The discussion below forcuses on some forms and uses for each of these. The inclusion or exclusion of particular recovery methods and/or possible products are arbitrary choices as a means of illustrating the scope of the problem and the progress to date in developing specifications.

There is no discussion of compost. Despite the widespread practice of composting municipal wastes in some parts of the world, composting is a well-documented economic failure in the United States, at least for large municipal plants [2]. Further, if some sort of broadly applicable specification were drawn for compost, it is unlikely it would require the breadth or detail required for other materials.

LIMITATIONS OF SPECIFICATIONS

At the early part of the century, Benjamin Dudley, who was president of the American Society for Testing and Materials from 1902 to 1909, described common faults in specifications [6]. These apply to the development of specifications for products recovered from wastes as well as from other materials. His teachings are worth repeating:

> The special fault which characterizes many specifications is the attempt on the part of the one who draws the specification to make it a place to show how much he knows. We have seen specifications which were apparently drawn with no other thought in mind than to embody all the knowledge the writer had on the subject. No discussion is needed on this point. The folly of it is apparent to all.

> Another fault is putting too many restrictions into the specification. According to our views, the fewer possible restrictions that a specification can contain, and at the same time afford the necessary protection in regard to the quality of the materials, the better the specification is. In some of our specifications, we have only one test; in others, perhaps half a dozen; the effort, however, being always to have the minimum number which will yield the product that is required.

> A third fault in specifications is in making the limits too severe. Some writers who draw specifications apparently put themselves in a position of absolute antagonism to those who are to make the material, and seem to have as a permanent thought in their minds to tie them down to the extreme limit. The maximum that a single

test piece shows, the minimum of an objectionable constituent
that may be obtained by analysis, the extreme point in elongation
that by chance some good and exceptional sample gives, are made
to represent the total output of the works. It is, perhaps, need-
less to say that such extreme figures are the worst possible mis-
take in making specifications.

Perhaps Dudley's counsel is easily illustrated by a word in modern English usage, "shoddy." The technical meaning of this word is
wool fiber recovered by the mechanical processing of old, wool fabrics.
The definition probably antecedes the Industrial Revolution in referring
to material recovered from wastes. Yet in modern usage "shoddy"
refers to anything of inferior worth or quality. With proper specifications, recovered materials need not be "shoddy."

One way of attempting to assure that specifications do not succumb
to the faults described by Dudley, and to assure that the specifications
are not arbitrary or exclusionary, is to develop and accept the specifications through a consensus process, involving those active in the
field. This is the practice of the American Society for Testing and
Materials and other voluntary consensus standards-setting organizations.

TYPES OF SPECIFICATIONS

As a broad generalization and simplification, specifications for materials may be categorized as describing performance, origin, or composition. Performance specifications are used in which the material
must meet mechanical, electrical, or chemical criteria, such as for
strength, insulating ability, or corrosion resistance. This type of
specification is better suited to finished products, such as iron pipe
or steel girder, than to the base materials. Performance specifications do not seem generally applicable to products recovered from
municipal wastes.

Origin specifications are used by the secondary materials industry and are illustrated by designations such as "genuine babbitt-lined
brass bushings" or "new zinc clippings" or "overissue news." The
origin specification is not only a description of the material, but also
often delineates its source and limits of principle contaminants. This
type of specification is essential for home and prompt scraps. It is
also useful for materials recovered from municipal wastes under certain circumstances, such as from voluntary recycling, hand separation, and perhaps from rejects of production of consumer items. This
is reflected in the published origin specifications titled "Shredded Tin

Cans" or "Old Can Stock" [7]. The origin specification for "incinerator bundles" describes the origin of a recovered steel product from an
established disposal process [8].

The focus of this chapter is on materials and fuels derived from
the mechanical and/or chemical processing of MSW. As such, origin
specifications will not always be useful. The reason is that the processing techniques are varied; many have been proposed, and new ones
are likely to be developed. At the present time, many new concepts
and processes are largely untried. Thus, there is no large body of
experience to make origin-type specifications meaningful. Separation
and recovery processes which have been proposed, or which are now
just operating, include steps which may not achieve complete separation and/or which may introduce contaminants. The existing origin-
type specifications do not make allowance for such possible contaminants and therefore cannot be used to assure quality and purity to
buyers.

DEVELOPMENT OF SPECIFICATIONS
FOR RECOVERED MATERIALS

Paper

Paper may be recovered from mixed municipal wastes by householder
separation or by mechanical processing, such as described in Chapter
4. United States industry currently accepts or rejects waste newspaper or corrugated board according to long-standing specifications
for such materials [9]. These specifications refer to paper furnishes
recovered by manual means only. By way of illustration of such specifications, Table 9.2 lists the requirements for the bottom (poorest)
and top (cleanest) grades of old newsprint. Obviously, these specifications will not be applicable to cellulose fiber recovered by wet-
pulping. It seems likely this product will have to match, in some way,
the specifications for merchant pulp. However, fiber recovered by
dry mechanical methods, such as described in Chapter 4, may ultimately be sold according to specifications resembling those in Table
9.2.

A possible way of recovering usable newsprint, and perhaps corrugated board, may be to handpick such materials from a conveyor
belt carrying the mixed waste from the tipping area of a resource recovery plant. This method of recovering commingled paper, adopted
from waste management practices of perhaps 50 to 80 years ago, was
described in 1972 [10] and has been practiced at the resource recovery plant operating in Milwaukee, Wisconsin. In order to recover

TABLE 9.2. Representative Origin Specifications for Old Newsprint

---

No. 1 news

Consists of newspaper packed in bales of not less than 54 inches in
    length, containing less than 5 percent of other papers.
        Prohibitive materials may not exceed. . . . $\frac{1}{2}$ of 1%
        Total outthrows may not excees 2%

Overissue news

Consists of unused overrun regular newspapers printed on newsprint,
    baled or securely tied in bundles, and shall contain not more than
    the normal percentage of rotogravure and colored sections.
        Prohibitive materials . . . . . . None permitted
        Total outthrows. . . . . None permitted

A prohibitive material is any included in a bale which if found in
greater than the specified level would damage the papermaking equipment and/or destroy the quality of the final product. Examples are
latex adhesives, magnetic inks, plastics, or asphaltic papers.

Outthrows are contaminants which make the product unsuitable for
consumption at the grade specified. Outthrows usually consist of
materials which are compatible with the papermaking process but if
found in amounts greater than the specified level will significantly
degrade the quality of the final product. Examples are cloth bindings,
chipboard, string bindings, and glassine.

Source: Ref. 9.

the newsprint this way, householders were urged to bundle and tie piles
of newspaper which were then collected commingled with other refuse.
Pickers were asked to select only those bundles which obviously were
not dirty. This method of collecting commingled paper does not add to
the cost of collection, as might other methods for separately collected
newspaper bundles.
    This method of recovering paper, and the cleanliness of the prod-
uct, were examined in actual trials. Some of the results have been
reported [11], and it is worth making additional observations.
    It was relatively easy to distinguish those bundles which were
obviously dirty after being commingled. Of course, this says little of
the dirt or other contamination which may be on the seemingly clean

newspaper. Some householders followed instructions about tying bundles with enthusiasm, using all sorts of old clothing and binding tapes
to secure the newspaper. Many of these bundles were not picked;
much of the tying materials could be classified as prohibitives or outthrows, as defined in Table 9.2. Some of the bundles tied with string
broke apart. The string on others broke but the bundles were more
or less intact and the paper could be picked. This resulted in the
observation that the bundles may best be recovered if they were not
tied, but placed inside kraft paper bags. This would enable a picker
to remove the contents and leave behind the soiled bag.

Some of the mechanical properties of the recovered paper were
measured (tear and tensile strength in both the machine and cross
directions) as possible indicators of degradation. However, the results of these measurements could not be distinguished from similar
values for control samples.

Paper recovered after being commingled with other waste, as
well as newsprint from separate collection and scrap from a printing
plant, were analyzed for moisture, content of water-soluble materials,
content of organic-solvent-soluble (fatty) materials, and content of
ash as indicators of "dirt." The results are summarized in Table 9.3
and show that the samples of commingled paper were not any more
contaminated than the separately collected or unused controls.

TABLE 9.3. Analysis of Recovered Newspaper

Property/sample Designation	Separately collected	Commingled	Control unused
Moisture (wt. percent, as received)	7.3	8.6 (48.4 on a rainy day)	9.1
Water solubles (wt. percent)	1.0	0.75	1.2
Organic solubles (wt. percent)	1.3	2.0	1.6
Ash (wt. percent)	0.47	0.79	0.50

Source: Ref. 11.

TABLE 9.4. Microbiological Content of Recovered Newspaper (Counts per sq cm)

| | Sample Designation[a] | | | | | |
	A	B	C	D	E	F
Total microorganisms	2.75	4.50	520	180	0.50	1.25
Bacterial spores	1.25	1.75	6.75	2.50	0	0.25
Total "cellulolytic" microorganisms	1.50	2.75	510	180	0.50	1.00
Fungi	0.25	0	13.3	160	0	0

[a] A—separately collected; B—separately collected but known to have been outside overnight; C—commingled, collected on a dry day; D—commingled, collected on a wet (rainy) day; E—commingled; F—unprinted newsprint scrap from a printing plant.
Source: Ref. 11.

The samples of commingled newspaper were analyzed for possible microbiological contamination, and the results are summarized in Table 9.4. These show that commingled paper may be contaminated as much, or as little, as separately collected paper. What cannot be shown, however, is the comparison with waste papers now being used in paper mills. This information is unavailable.

The investigation of the properties of commingled paper is an essential starting point in order to provide the basis necessary for the development of a specification. No such specification is yet available.

Steel

The preponderant form of steel in mixed municipal waste (from mostly household sources) is cans, both tinplate and tin-free steel. Other magnetic metals discarded as obsolete scrap (e.g., appliances and autos) are usually separately processed and recycled.

At least in the United States, the composition of recovered magnetic metals varies slightly from one locale to another, although the basic components are usually the same. Assays in several cities

showed that in the household portion of waste, some 80 to 90 percent
of the magnetic metals were cans of various types [12].

Cans are not all magnetic metal; only from 86 to 96 percent by
weight of cans is steel, depending on construction [13]. The remainder is aluminum, tin, side-seam solder, and coatings. These materials must be taken into account in developing specifications for recovered cans.

There are several possible uses or markets for recovered steel
cans [14, 15], including substitutes for iron ore as a charge to blast
furnaces; charge for steelmaking in the open hearth, BOF, Q-BOP, or
electric arc furnaces; remelt in electric furances; chemical detinning;
raw material for ferroalloy production; and use as a precipitation iron
in the recovery of copper from leaching ores. Each of these uses has
been addressed by the American Society for Testing and Materials
(ASTM) Committee E-38. Table 9.5 summarizes salient points of the
standards. This table reflects the results of deliberations and ballots
of the committee, and so is a consensus viewpoint of the participants
from the resource recovery and user industries.

As can be seen in Table 9.5, current United States practice is to
discourage any steel scrap which has been incinerated or otherwise
burned. The reasons for this include high ash, oxidation, and alloying

TABLE 9.5. Tentative Composition Limits for Recovered Steel[*]

Intended use	Total combustibles (wt. %)	Bulk density (kg/m^3)	Physical form	Chemical comp. limits
Cu industry	0.2	480	Loose, not baled or balled	None
Foundries	4.0	800	Loose, baled or balled	Note 2
Iron and steel production	4.0	1200	Baled or loose	Note 2
Detinning	Note 1	400	12.5 X 152 mm	Note 1
Ferroalloy production	0.5	800	--	Note 3

*ASTM Standards E700 and E701.

TABLE 9.5.   (Continued)

Note 1: Chemical requirements for detinning

Component	Composition (%)
Al	4. 0 (not based on melt analysis)
Sn	0. 15

Note 2: Chemical composition limits

Elemental limitation	Iron and steel (wt. %)[a]	Foundries (wt. %)
P	0.03	0.03
S	0.04	0.04
Si	0.10	—
Ni	0.08	0.12
Cr	0.10	0.15
Mo	0.025	0.04
Cu	0.10	0.20
Al	0.15	0.15
Sn	0.30	0.30[b]
Pb	0.15	0.03
Zn	0.06	0.06
Yield:	90	90

[a]Experience has shown that material which has been incinerated will probably not meet these requirements.
[b]0. 10 maximum for steel castings.

Note 3: Chemical requirements for ferroalloy production

Component	Maximum content (wt. %)
C	0. 06
Mn	0. 35
P	0. 03
Cr	0. 15
Cu	0. 20
Al	0. 15
Sn	0. 30
Ti	0. 025
Yield:	90

with tramp elements. Recently, it was reported that incinerated scrap
can be used to manufacture rolled sheet, if only on a pilot scale of 900-
kg melts [16]. It is impossible to say if this research will lead to
changes in the specifications to permit incinerated scrap steel cans.

An ASTM Standard has also been developed for test methods to
accompany the specifications, such as for determination of bulk density and content of tramp organic contaminants. The two standards
are designated E700 and E701.

It is worth noting that the specifications for reuse of recovered
steel permits inclusion of tin-plated metal. The presence of tin, in
the amounts usually found on cans, can be tolerated in many end uses
and the amount of tin which will accumulate through repeated recycling of tin-plated scrap can be tolerated in several new steel products
[17]. Similarly, a certain amount of aluminum from bimetal cans and
organic contaminants are permitted for several end uses. The effect
and tolerance of these contaminants has been discussed [18].

Table 9.6 is a representative origin specification for steel scrap.
Comparison of Tables 9.5 and 9.6 gives some indication of the development of composition specifications for recovered materials from
wastes as compared to the origin specifications of the scrap industry.

TABLE 9.6. Representative Origin Specifications for Steel Scrap

---

Shredded tin cans for remelting

Shredded steel cans, tin-coated or tin-free, may include aluminum
tops but must be free of aluminum cans, nonferrous metals except
those used in can construction, and nonmetallics of any kind.

Bundled no. 2 steel

Wrought iron or steel scrap, black or galvanized, 1/8 inch and over
in thickness, compressed to charging-box size and weighing not less
than 75 pounds per cubic foot. Auto body and fender stock, burnt or
hand-stripped, may constitute a maximum of 60 percent by weight.
(This percent based on makeup of auto body, chassis, driveshafts, and
bumpers.) Free of all coated material, except as found on automobiles.

Source: Institute of Scrap Iron and Steel, Inc., Washington, D. C.

Aluminum

Aluminum discards in municipal solid waste are small as a fraction of the total mass, but valuable as a fraction of the total revenue which could be obtained by a resource recovery plant [19]. Further, the recovery and reuse of scrap aluminum conserves a considerable amount of energy compared to manufacture of new metal from ore, as discussed in Chapter 2.

The aluminum discards are likely to be made from many of the more than 225 Aluminum Association registered alloys. Further, in parts of the United States, this will include the wrought alloys used in the manufacture of beverage cans. Most of this scrap, however, will be the common 1100, 3000, and 5000 series alloys, most of which can be blended to make 3105 and 3004 alloys [20]. Some of the cast products will contain high silicon, copper, and zinc which are tramp to the manufacture of 3105 and 3004.

Secondary smelters will blend various sources of scrap and produce, in the main, casting alloys such as 380, relatively high in content of silicon, copper, and zinc. Beverage container alloys (3004 and 5182) are high in manganese and magnesium and low in silicon, copper, and zinc so that, broadly speaking, they are not compatible with casting alloys. The converse is also true; casting alloys generally cannot be included with recovered metal for manufacture of new 3004 alloy [20].

Recovered alluminum must be free of sand, grit, and particularly glass. At the melt temperature, aluminum will reduce the silica (SiO_2) in the glass to silicon (Si), which will alloy with the aluminum and possibly raise the silicon level above that which can be accepted for wrought products. The content of iron is similarly a problem and must be kept low during recovery, as it is difficult to reduce by subsequent metallurgical processing.

Organic contaminants in the recovered metal will burn off in the furnace and either cause an additional load on the air pollution control equipment or be included in the dross (slag). In either case, organic materials are a contaminant in that they add to processing costs and reduce the recovery of metal. Specifications must also include a requirement on piece size and absence of fines (small particulate material), so as to control metal losses due to oxidation during melting. Many of these points have been discussed [20, 21].

There is a clear hierarchy of uses of recovered aluminum, starting with the highest value reuse in the manufacture of new 3004 wrought sheet and descending to the lowest value for casting and "de-ox" grades. The specifications evolving from the ASTM committee reflect this hierarchy of use, value, and consequent narrowness of chemical specifica-

tions. Some six grades of recovered aluminum have been identified.
Their chemical composition limits are given in Table 9.7.

The cognizant Non-ferrous Metals Subcommittee of ASTM E-38 is
also developing test methods to accompany the specifications; publication is expected in 1980. What are required are methods to determine
the weight fraction of fines (material smaller than 12 U. S. mesh,
1.68 mm) and the melt recovery. There are two schools of thought
regarding the latter: that it should be determined in a production
furnace, or that it should be determined by laboratory assay. Several
methods are already available and in use for the chemical analysis of
the metal, such as ASTM methods E-101 and E-227.

Judging from reported experiences, it is likely that the various
methods which have been proposed for aluminum recovery can produce
aluminum scrap that will meet the chemical composition limits for
Grades 1 and 2 [21, 22].

The composition-type specifications of Table 9.7 are excellent examples of the evolution of specifications for recovered materials from
origin-type specifications. By way of contrast, the origin specification for "Old Can Stock" [7] reads, in total, "Shall consist of clean
old aluminum cans decorated or clear, free of iron, dirt, liquid and/
or other foreign contamination." This specification makes no allowance for hierarchy of uses (hence downgrade) nor recognizes that the
recovery method may produce metal with organic contaminants, say
from eddy current separation [22] or iron and silicon contamination,
say from heavy media separation [23].

Glass

Recovered glass can be reused in the manufacture of glass containers,
in such secondary products as building materials, or as aggregate.
The requirements for each use differ widely.

In glass container manufacturing, the raw materials must be carefully controlled so as to maintain batch-to-batch reproducibility [24].
The feedstock or "batch" to the glass furnace must be free of refractory particles which will not melt; these are the cause of "stones" in
the glass (bits of ceramic materials, which are unsightly and as possible points of stress concentration may cause weakness and bottle
breakage). Bubbles in the glass—known as "seeds"—also produce an
undesirable appearance and can cause weakness. Seeds can be caused
by inclusion of organic materials (which will decompose) in the charge
to the furnace. Recovered glass, if not processed properly, can cause
both stones and seeds. For example, aluminum is a potential source

TABLE 9.7. Tentative Chemical Composition Limits for Recovered Aluminum (wt. %)

Element[a]	Grade 1	Grade 2	Grade 3	Grade 4	Grade 5	Grade 6
Si	0.30	0.30	0.50	1.00	9.00	9.00
Fe	0.60	0.70	1.00	1.00	0.80	1.00
Cu	0.25	0.40	1.00	2.00	3.00	4.00
Mn	1.25	1.50	1.50	1.50	0.60	0.80
Mg	2.00	2.00	2.00	2.00	2.00	2.00
Cr	0.05	0.10	0.30	0.30	0.30	0.30
Ni	0.04	0.04	0.30	0.30	0.30	0.30
Zn	0.25	0.25	1.00	2.00	1.00	3.00
Pb	0.02	0.04	0.30	0.50	0.10	0.25
Sn	0.02	0.04	0.30	0.30	0.10	0.25
Bi	0.02	0.04	0.30	0.30	0.10	0.25
Ti	0.05	0.05	0.05	0.08	0.10	0.10
Others, each:	0.04	0.05	0.05	0.08	0.10	0.10
Others, total:	0.12	0.15	0.15	0.20	0.30	0.30
Al	Bal	Bal	Bal	Bal	Bal	Bal

[a]By agreement between purchaser and producer, analysis may be required and limits established for elements or compounds not specified above.

Source: ASTM Subcommittee E-38.03, November 1977.

of stones. At glass melt temperatures in the absence of air, aluminum reduces silica in glass to silicon, which is refractory and forms stones [24].

Glass must also be free of ferrous materials which will change the glass color. Amber and green glass contain varying amounts of iron and chromium compounds which will affect the color of the final glass when added as cullet.

Reportedly, from 15 to 20 percent cullet by weight has been used in glassmaking routinely [25]. However, it has been shown that more cullet—maybe as much as 100 percent—can be used in production melts under appropriate circumstances [24].

Several colors of glass are present in municipal solid waste in the United States. About two-thirds of the glass produced for containers is flint (clear). Georgia green is used for the familiar Coca-Cola bottle; other soft drink bottles are the deeper emerald green; champagne green is used for the deep green wine bottles; amber glass is used largely for beer bottles. Also, when refuse is sampled, there are always minute amounts of blue, opal, and other colors of glass.

The mix between amber and green seems to depend on the part of the country in the United States, with more green on the West Coast than the East. A materials separation plant operating on mixed municipal waste may choose to separate flint (containing some Georgia green), leaving behind a mixture of amber and green [26]. The flint fraction can be used to make new flint or other containers; the color-mixed fraction can be used to make amber or green containers [27].

Thus far, there is not enough experience in using glass for building products to be able to write specifications. Some of the uses apparently do not require a clean product [28]. Use in brickmaking as a flux may require specifications for particle size and organic content [29].

In the past, manufacturers commonly called for cullet which was "clean, color-sorted, and free of metals." It was relatively easy to follow these instructions because much of the cullet was obtained from brewers and bottlers by cullet dealers, some of whom processed or beneficiated the cullet to meet customer requirements. Increased labor costs overpriced these and similar sources so that less cullet was purchased.

In order to provide guidance for development of separation systems for municipal wastes, the Glass Packaging Institute developed guideline information to describe cullet quality acceptable for use in glass-container-manufacturing furnaces and reflect manufacturing experience. The bases for the specification have been described [24]. The outstanding feature of this specification was the requirement for extremely low levels of "stones" or refractory particles. For example, cullet particles larger than 20 U.S. mesh (0.9 mm) could not contain more than one refractory particle in 18 kg. In some European countries, glass container manufacturers list this requirement as "none" and so fall prey to some of the faults described by Dudley.

Glass may also be recovered from municipal wastes by a process called "froth flotation." This uses technology from the mineral-dressing industry to produce a clean, sandlike, color-mixed cullet with pieces approximately between 0.09 and 0.9 mm (140 × 20 U.S. mesh) in size. While application of the froth flotation technology to resource recovery is just beginning [30], early indications are that the material produced will be clean and acceptable for the manufacture of glass containers.

Table 9.8 is a draft version of the specification for froth floated glass considered by the ASTM Resource Recovery Committee. Because the particles are smaller than would be produced by color-sorting, the requirement for stones (refractory material) is considerably relaxed compared to 1 particle in 18 kg. Material meeting this specification is intended as a substitute for 20 percent of the charge to a glassmaking furnace, although this substitution will be optional with the furnace operator.

The cognizant subcommittee has also developed test methods to be used in association with the specification of Table 9.8, ASTM Standard E688. Additional methods for sampling shipments must also be developed.

Refuse-Derived Fuels

The development of specifications for refuse-derived fuel (RDF) is the subject of much current research. The specifications are dependent on the particular conversion and use technologies, as described in Chapter 5. Also, the specifications must be closely coupled with development of methods for testing and sampling.

A fuel is normally described by its proximate and ultimate analyses. The ultimate analysis of RDF will reflect its major constituent, cellulose, and the minor content of such things as plastic films. The proximate analysis will depend greatly on what processing steps are used to reduce the ash, such as screening to remove inorganic fines (as discussed in Chapter 5) and possibly drying (as discussed in Chapter 8).

The development of specifications for RDF is being preceded by development of associated sampling and test methods, particularly for sieve size, moisture, and proximate and ultimate analyses. This work is being conducted principally by Subcommittee E-38.01 on Energy of

TABLE 9.8. Tentative Composition Limits for Waste Glass as a Raw
Material for the Manufacture of Glass Containers

---

1. A preponderant portion of glass cullet will be soda-lime bottle
    glass containing: SiO_2 66-75, Al_2O_3 1-7, CaO + MgO 9-13, and
    Na_2O 12-16 weight percent.

2. The sample shall show no drainage and be noncaking and free-
    flowing. A moisture content of less than 0.5 wt. % is required to
    meet the free-flowing characteristics of cullet which is predom-
    inantly of smaller particle size (<1.18 mm, or smaller).

3. Particle size: 100% <50 mm, <15 wt. %, <0.11 mm (140 U.S.
    mesh).

4. Organic materials: <0.2 wt. %.

5. Magnetic metals: <0.14 wt. % with no particle >6 mm.

6. Other inorganic materials: In the particle size range > 0.85 mm
    (20 U.S. mesh), <0.1 wt. %. For particles in the size range
    smaller than this, <0.5 wt. %. No particles >6 mm.

7. Refractory materials: The following limits shall apply with no
    particles > 6 mm:

Particle size (p mm)	No. of particles permitted
p > 0.85 (>20 U.S. mesh)	1 per 18 kg of sample
0.85 > p > 0.43 (20 × 40 U.S. mesh)	2 per 450 g of sample
0.43 > p > 0.25 (40 × 60 U.S. mesh)	20 per 450 g of sample

8. Nonmagnetic metals: All particles must be smaller than 6 mm.
    Particles larger than 0.9 mm (20 U.S. mesh), no more than 1
    particle per 18 kg.

the ASTM Resource Recovery Committee E-38. A brief progress
report was published [5].

Specifications for steam or electricity are the same whether these
products are generated from waste or fossil fuels. Future specifications for solid refuse-derived fuels will no doubt resemble similar
specifications for coal. A caution is that the specifications for coals
have evolved over many years and points of specification and analysis
may be included for historical or technical reasons not related to RDF.
There is a danger that drafters of specifications for RDF may unin-
tentionally and indiscriminately include items which are not germane.

On the other hand, there are points which might be included in the
specifications and test methods for RDF which are not germane to
coals. This is especially true in the adaptation of methods of analysis
so they are more applicable to the nature of RDF.

The calorific value of either coal or RDF is determined on a small
sample of a large shipment. The measurement is made in a laboratory
calorimeter. Briefly, the method used for sampling large quantities
of coal is to take a gross sample of perhaps several hundred pounds
and divide this into a small (few-gram) sample in a prescribed manner.
The final sample is not necessarily representative of the large amount
of coal, but buyers and sellers have agreed over the years to accept
this method in the absence of a better one. To date, probably because
it is a new field, buyers and sellers have not yet agreed on any similar
method for sampling RDF. They will undoubtedly have to; it is impossible to obtain a representative sample of fuel from thousands of tons
without an inordinate amount of effort.

One proposal for overcoming the concerns of unrepresentative
samples is to charge larger quantities of material into the laboratory
calorimeter. Although this view is well meaning, it does not fully
recognize the difficulties in designing and operating larger calorimeters
of the type needed (so-called bomb calorimeters). In addition, there
is the argument that if the calorimeter were capable of accepting, say,
a 1-kg sample instead of the current 1- to 2-g sample, the 1000-fold
increase would not necessarily provide a more representative sample.
In addition, if unprocessed refuse were being analyzed, even a 1-kg
sample would have to be shredded or otherwise homogenized and could
not easily accommodate bulky items in the waste.

It appears that the precision of calorific analyses is about 10 percent coefficient of variation or less. (The coefficient of variation is
the standard deviation of a series of measurements, divided by the
mean, multiplied by 100.) The question still has to be resolved if this
is good enough or if a greater precision is necessary. It is really a
decision to be made mutually by buyers and sellers of the fuel.

There are several sources of error in performing fuel analyses of any sort. These include systematic errors in the methods of measurement (no method is perfect), sampling (nonrepresentative aliquots), and the normal variation in the fuel. The latter may be quite large in RDF produced by a variety of technologies. In short, it may not be possible realistically to expect coefficients of variation much less than about 10 percent.

FUTURE PROSPECTS

There are many activities and many people involved in the development of specifications and related test methods. A major focus of this activity is Committee E-38, Resource Recovery, of the American Society for Testing and Materials [31]. This committee, like all within ASTM, has a balanced membership among producers and users of recovered products and general interests in the field. ASTM is a voluntary consensus standards-setting organization. In all likelihood, the consensus process will produce many needed specifications and test methods. Although these standards are voluntary, municipalities and others may elect to include the standards in contractual documents for waste-to-energy plants.

NOTES AND REFERENCES

1. Although the quantity of municipal waste generated and disposed of each year seems enormous, the corresponding quantity of recovered materials is apt to be small (at least in the foreseeable future) compared to production of new materials which could utilize scrap. As an example in the United States, the MSW in the standard metropolitan statistical areas totals approximately 100×10^6 tons per year, and perhaps 6 or 7 percent of this is magnetic metal. If all of this metal were recovered for reuse in the manufacture of new iron and steel, it would represent only about 5 percent or so of new steel production. It is unlikely that this small fraction could be enough to force large technological change at major steel-producing mills. This example may be oversimplified, especially considering that large quantities of scrap are now used to manufacture "new" steel. However, the point should be clear.
2. G. E. Stone, C. Wiles, and C. Clemons. Composting at Johnson City. Final Report on Joint USEPA-TVA Composting Project. Vols. I and II. Report EPA/530/SW-31r.2 Environmental Protection

Agency, Cincinnati, 1975. Also, in 1971, of 18 composting plants constructed in the United States only three were operating on a commercial basis: Guidelines for local government on solid waste management. Environmental Protection Agency, Washington, D.C., 1971.

3. H. Alter and W. R. Reeves. Specifications for materials recovered from municipal refuse. Report EPA-670/2-75-034. Environmental Protection Agency, Cincinnati, 1975.

4. P. M. Sullivan and H. V. Makar. Quality of products from Bureau of Mines resource recovery systems and suitability for recycling. In Proceedings, Fifth Mineral Waste Utilization Symposium (E. Aleshin, ed.). IIT Res. Inst. and Bureau of Mines, Chicago, 1976, pp. 223-33.

5. H. Alter. Development of specifications for recycled products. Conservation and Recycling 2(1): 71-84 (1978).

6. ASTM. The life and work of Charles Benjamin Dudley. Philadelphia, 1911, pp. 127-28.

7. National Association of Recycling Industries. Standard classification for nonferrous scrap metals. Circular NF-73. New York, 1973.

8. Institute of Scrap Iron & Steel. Accepted specifications for selected remelting grades of steel scrap. Washington, D.C., 1973.

9. Paper Stock Institute of America. Paper stock standards and practices. Circular PS-75. New York, 1975.

10. National Center for Resource Recovery. Materials recovery system: Engineering feasibility study. Washington, D.C., 1972.

11. H. Alter, K. L. Woodruff, A. Fookson, and B. Rogers. Analysis of newsprint recovered from mixed municipal waste. Resource Recovery and Conservation 2: 79-84 (1976).

12. H. Alter, S. L. Natof, K. L. Woodruff, and R. D. Hagen. The recovery of magnetic metals from municipal solid waste. Report RM 77-1. National Center for Resource Recovery, Washington, D.C., 1977.

13. E. J. Ostrowski. Recycling of tin-free steel cans and scrap from municipal incinerator residue. Presented at the 79th General Meeting, American Iron and Steel Institute, New York, May 26, 1971.

14. C. Lipsett. Industrial wastes and salvage, 2nd ed. Atlas, New York, 1963.

15. A. Darnay and W. Franklin. Salvage markets for materials in solid wastes. Report SW-29c. Environmental Protection Agency, Washington, D.C., 1972.

16. R. S. Kaplan and H. V. Makar. Use of refuse magnetic fractions for steelmaking. Resource Recovery and Conservation 3: 69-96 (1978).

17. E. J. Duckett. The influence of tin content on the reuse of magnetic metals recovered from municipal solid waste. Resource Recovery and Conservation 2: 301-28 (1977).

18. E. J. Duckett. Contaminants of magnetic metals recovered from municipal solid waste. Report NSF/RA-770244. National Center for Resource Recovery, Washington, D. C. , 1977.

19. J. G. Abert, H. Alter, and J. F. Bernheisel. The economics of resource recovery from municipal solid waste. Science 183: 1052-58 (1974).

20. S. T. Abbate. Use of aluminum recovered from municipal solid waste. In Resource recovery and utilization (H. Alter and E. Horowitz, eds.). ASTM, Philadelphia, 1975, pp. 106-13.

21. G. F. Bourcier and K. H. Dale. The technology and economics of the recovery of aluminum from municipal solid waste. Resource Recovery and Conservation 3: 1-18 (1978).

22. H. Alter, S. L. Natof, and L. C. Blayden. Pilot studies processing MSW and recovery of aluminum using an eddy current separator. In Proceedings, Fifth Mineral Waste Utilization Symposium (E. Aleshin, ed.). IIT Res. Inst. and Bureau of Mines, Chicago, 1976, pp. 161-68.

23. E. L. Michaels, K. L. Woodruff, W. L. Freyberger, and H. Alter. Heavy media separation of aluminum from municipal solid waste. Soc. Mining Engrs. AIME Transactions 258: 34-40 (1975).

24. C. E. Seeley. Glass in solid waste recovery systems. In Resource recovery and utilization (H. Alter and E. Horowitz, eds.). ASTM, Philadelphia, 1975, pp. 114-21.

25. Kirk-Othmer. Encyclopedia of chemical technology, 2nd ed. Wiley, New York, 1966, 10: 533-604.

26. J. Cummings. Glass and aluminum subsystem—Franklin, Ohio. In Proceedings, Fourth Mineral Waste Utilization Symposium (E. Aleshin, ed.). IIT Res. Inst. and Bureau of Mines, Chicago, 1974, pp. 104-15.

27. P. Scott. Report on glass workshop. In Resource recovery and utilization (H. Alter and E. Horowitz, eds.). ASTM, Philadelphia, 1975, pp. 159-64.

28. Utilization of waste glass in secondary products. In Proceedings of the conference at Technology Applications Center. University of New Mexico, Albuquerque, 1973, pp.

29. M. Tyrell and A. Goode. Waste glass as a flux for brick clays. Publication RI 7605. Bureau of Mines, Washington, D. C. , 1972.

30. B. Morey. Glass recovery from municipal trash by froth flota-
     tion. In Proceedings, Third Mineral Waste Utilization Symposium
     (M. A. Schwartz, ed.). IIT Res. Inst. and Bureau of Mines,
     Chicago, 1972, pp. 311-22; H. Heginbotham. Paper presented at
     the Sixth Mineral Waste Symposium, Chicago, 1978.
31. American Society for Testing and Materials, 1916 Race Street,
     Philadelphia, PA. 19103.

Chapter 10

OVERVIEW AND IMPLEMENTATION

The preceding chapters centered on the technical and economic aspects of waste-to-energy systems and materials recovery practices in the United States and Europe. The following discussion presents observations based on this information to assist municipal decision makers in judging the applicability and feasibility of implementing energy and materials recovery in their communities.

Preserving public health is the primary function of solid waste management. Resource recovery is therefore an adjunctive dimension and should be practiced only when it makes economic and technical sense. Resource recovery has the potential for lowering costs and reducing adverse environmental effects associated with disposal. In addition, it is a method of materials and energy conservation.

The resource recovery method (or methods) adopted by a community will be largely determined by the amount and composition of the waste available for processing, the technical requirements, and the capacity of the local markets to use the recovered products. Resource recovery does not replace disposal; not all waste is recoverable, and a community will need a landfill for residues. As such, resource recovery processing does not require the same redundancy and reliability of a disposal method essential to maintaining public health. Resource recovery systems require relatively large investments, so redundancy is expensive. A fraction of the cost of redundant systems may be better spent on a contingent landfill.

Resource recovery systems may be best planned to process a portion rather than all waste generated by a community. It is difficult to estimate the amount of waste to be generated over the amortized life of the plant. Furthermore, processing only a portion hedges against the risks associated with the comparatively new technology and practice of resource recovery.

BACKGROUND, COMPOSITION, AND QUANTITIES

The history of electric power use shows that the United States is the highest per caput consumer in the world. This is likely to continue despite conservation as long as the country strives for the world's highest standard of living. The industrialized world recognizes the effects of pollution control and dwindling fossil fuel supplies on life styles and energy consumption habits. These factors, coupled with solid waste disposal problems in urban areas, have focused attention on recovering energy from solid waste. It is likely that resource recovery from waste will be normal practice in the future.

Europe has used solid waste for energy generation longer than the United States. There are many reasons, including a generally higher cost for solid waste disposal in Europe. In part, the higher prices are related to the lack of economies of scale of smaller populations, less waste, and generally less land to use for disposal. Over the past several decades, European solid waste has also contained more moisture and food waste as well as less recoverable metals and glass, factors related to different practices of preparing and distributing food. However, these contrasts are increasingly less apparent.

Many European cities are close to agricultural areas, which has enabled some of them to find markets for compost prepared from waste. The more common practice has been to reduce waste volumes by incineration to avoid devoting land areas to disposal. Heat recovery incinerators are used since many cities are responsible for waste disposal and electricity and/or steam generation. As European cities seek to improve air quality by tightening their air pollution control laws, the cost of incineration is increasing.

In Europe, the heat recovery incinerators are frequently used to dispose of sewage sludge. United States municipal decision makers may shudder at the institutional problems likely to arise by combining these functions. European cities have many of the same problems, but they have been coping with them longer.

The economies of scale of resource recovery in the United States and Europe differ because of contrasting populations, waste generation rates, and waste compositions. There are more recoverable metals in United States waste, thus providing a larger source of revenue. Because of the differences in quantities and composition, resource recovery objectives in the United States and Europe are different. In the former, the objective is to recover and sell metals (and perhaps glass) and convert the balance to a refuse-derived fuel (RDF). In the latter, the emphasis is on paper recovery, some metal recovery, and mass burning. In years to come, the objectives will likely be more similar.

CONVERSION AND USE TECHNOLOGIES

Mass burning in heat recovery incinerators is widely practiced in Europe. There is a question if this may be considered a "proven" technology (as defined in Chapter 5). There are many examples of RDF production in the United States, and a larger question as to whether it is a "proven" technology.

The United States is moving toward recovery of ferrous and nonferrous metals and glass. In Europe, there is generally less ferrous metal and much less nonferrous metal in waste, and the interest in the recovery and reuse of glass for containers is low. (The low interest in glass recovery may change as United States plants gain experience.) There is greater interest in plastics recovery in Europe (a subject not addressed in previous chapters) than in the United States.

The technologies for use of recovered materials is similar for the two continents; the technology of steelmaking and papermaking has evolved with technological interchanges across the Atlantic. Therefore, the specifications and recovery technology for these materials should be the same. The disparate interests in implementing recovery technologies stem primarily from the contrasting waste compositions in various countries.

ECONOMIC MODEL FOR SYSTEMS SELECTION

Chapter 7 describes a systematic method for gathering and presenting information in planning resource recovery. The method is based on computing "how much" resource recovery a community can afford while keeping its cost of disposal constant.

An important step is knowing the exchange or floor prices likely to be received for the recovered materials and fuel. There are two reasons for this: (1) an early revenue forecast is an important part of the analysis; and (2) markets come first, and the specifications for the recovered materials and fuel are what determine which technology or system to implement.

Chapter 8 is a "how to" manual for the marketing of recovered products in advance of having a recovery plant. Some of the technical and institutional aspects of this advance marketing have been discussed. More important, such marketing has been achieved.

The planning and analysis methodology of Chapter 7 results in the computation of the indifference value of the fuel (or other energy form) which must be obtained to keep the cost of disposal constant. Computation of this value is an important starting point in any advance marketing effort.

IMPLEMENTING RESOURCE RECOVERY

The information presented can assist municipalities to implement resource recovery. Some procedural guidelines follow to assist the planner.

Step 1: Quantity and Composition. The quantity of waste available for resource recovery plants must be determined. Because of the difficulties of predicting waste volumes, a community may want to obtain a truck scale as a first step. It is unlikely the amount of waste will change in future years, except as population increases. If per caput generation rates are used, they must be adjusted downward to allow for commercial and industrial wastes attributed to per caput generation but not collected or included with household wastes.

The composition of the waste is even more difficult to determine Samples can be picked apart and analyzed, but this is time-consuming, expensive, and determines the composition only for the day of the analysis. Picking does not allow for changes caused by seasonal variation or changes in packaging and marketing practices which may occur in the future. The planner may be well advised to use average composition data (for that part of the country) and be prepared to adjust financial forecasts (and accompanying contract documents, if the plant is privately owned or operated) when the plant is operating and gains experience. One city contracted for a plant on this basis in 1975.

Step 2: Markets and Specifications. To secure markets, the planner may want to use the methods and format for the letter of intent (LoI) in Chapter 8.

For each recovered product or market, there will be an accompanying specification. The plant must be designed to meet specifications, the same as any other manufactured commodity.

In some areas, the local market for an energy product will be evident. For example, if there is a local utility burning coal in suspension (use technology C in Chapter 5), they should be approached first. If there is not an obvious customer for a solid fuel, then customers for the more "refined" energy products, such as steam or electricity, should be sought. Increasing the degree of "refinement" (solid fuel to steam to electricity, for an example) raises the selling price but increases investment and production costs.

Step 3: Determining the Technology. The specifications from step 2 provide the "musts" for the plant design. At this point, an engineering analysis is needed. It may be a preliminary design or just a comparison with literature descriptions. The technology should not be determined by using a request for proposals (or similar bid document). At this stage, it would be premature.

Step 4: Determining the Cost. The market LoI (to bid or buy) and the preliminary engineering analysis of step 3 estimate the plant revenues and costs. These are entered into an analysis model, such as in Chapter 7. At this point, the cost and type of financing should be selected or estimated.

Step 5: Determining the Financing. Resource recovery plants have been financed through both the public and private sectors. Many different kinds of debt instruments have been used. These include general obligation bonds, revenue bonds, pollution control bonds, and—in some rare instances—the private sector contractor paid for the plant. The method employed depends partially on local laws, restrictions, and practices.

Step 6: Determining a Method of Management. Resource recovery plants may be publicly or privately owned and operated, or in some combination. The pros and cons have been discussed. (See the publications in the Bibliography at the end of this chapter.) Nevertheless, resource recovery is an enterprise that produces marketable material or fuel to a specification. As such, it is different in concept and operation from some of the service functions usually performed by a municipality—fire, police, and social services. The method of management will partially determine the cost. In addition, the presumably higher costs of private management or ownership may be offset by greater efficiency.

Step 7: Site Location. The problems associated with the selection of sites for disposal facilities have also plagued resource recovery. Neighborhoods must be convinced that a resource recovery plant is an industrial facility and not a "dump."

Step 8: The Procurement. The results of the preceding seven steps must be combined with the usual legal requirements of public procurements and some sort of request for proposals must be issued. Perhaps the major stumbling block to implementing resource recovery has been poorly drawn-up requests for proposals which resulted in no procurement. Hopefully, the steps outlined here will lead to a successful procurement.

CONCLUSION

The information presented here is intended to assist public officials, in particular, better to understand resource recovery, to interpret information as it applies to their localities, and to proceed, if conditions warrant, with resource recovery in their communities.

Resource recovery, the reuse of materials and the conversion of waste to energy, will enhance national and global energy conservation.

Waste-processing ventures will involve risk, uncertainty, and obstacles. But like other similar endeavors, experience is a good teacher. The many waste-to-energy systems operating today show that it can be done.

It is hoped that this discussion of American and European experiences will increase mutual understanding and lead to greater conservation of material and energy resources on both continents.

BIBLIOGRAPHY

Abert, J. G. Procurement and contracting. In Resource recovery and utilization (H. Alter and E. Horowitz, eds.). ASTM, Philadelphia, 1975, pp. 31-9.
Abert, J. G. Resource recovery: The economics and the risks. Professional Engineer 45: 29-31 (1975).
Alter, H. Resource recovery cannot be dependent on subsidy. Solid Wastes Management/Resource Recovery 5: 54-55 (October 1974).
Alter, H. Role of the scrap processor in municipal solid waste resource recovery. Scrap Age 32(4): 213-18 (1975).
Alter, H. Pitfalls when planning resource recovery. Waste Age 7(3): 12-14 (1976).

Appendix A

SYMPOSIUM ON SOLID WASTE CONVERSION TO ENERGY
HAMBURG, GERMANY, SEPTEMBER 17, 1977

PROGRAM AND PARTICIPANTS:

<u>Presiding:</u> John Bergacker, President
International Federation of Municipal Engineers
Miami Beach, Florida

<u>Welcome and Introductory Remarks:</u>

Dr. Ing. Prof. Müller-Ibold
Chief of Building and Public Works
Hamburg, Germany

<u>Solid Wastes and Energy Trends in Perspective:</u>

Hans Wasmer, Director
WHO International Reference Center on Waste Disposal
Dubendorf, Switzerland

Alex Radin, Executive Director
American Public Power Association
Washington, D. C.

<u>Energy Recovery from Solid Wastes:</u>

Donald K. Walter
Chief, Urban Waste Technology Branch
Energy Research and Development Administration
Washington, D. C.

Commentators:

> L. Barniske, Diplomatic Engineer
> Specialist for Thermal Processes
> Environmental Department
> Berlin, West Germany
>
> Dr. Ir. B. G. Kreiter
> Stichting Verwijdering Afvalstoffen
> Amersfoort, Netherlands
>
> Jean Colardeau
> Public Works Engineer
> Chief, Bureau for Urban Wastes
> Ministry of the Environment
> Paris, France

Governmental Institutions vs. Public Service Corporations:

> Dr. Karl Wolf
> Professor of Management
> Social Technology and Economics
> Fachhoch Schule Nord Deutschland
> Vechta (Oldenvurg), West Germany

Presiding:

> J. G. Van Putten
> Secretary General
> International Union of Local Authorities
> The Hague, Netherlands

Economic Model for Selecting RDF Systems:

> James McCarty
> President, American Public Works Association, and
>   Director of Public Works
> Oakland, California

Presiding:

> Professor Eugenio de Fraja Frangipane, President
> International Solid Wastes and Public Cleansing Association
> Milan, Italy

Commentators:

> Jean Defeche
> Deputy Chief Controller
> Electricity of France
> Paris, France
>
> Derek Ayers, Director
> Department of Public Health Engineering
> Greater London Council
> London, England

Organization, Financing, and Management Options:
(Review of Practices in Four Countries)

> George Kerr
> Minister of the Environment
> Province of Ontario, Canada
>
> Herbert Oppermann
> Director of Public Cleansing
> Hamburg, Germany
>
> James Caplinger
> Director, European Office
> Council for International Urban Liaison
>
> Terry Sprenkel
> City Manager
> Ames, Iowa
>
> Marshall Suloway, Commissioner
> Department of Public Works
> Chicago, Illinois
>
> Svend Seitzberg
> Managing Director
> Solid Wastes Contracting Company
> Copenhagen, Denmark

Summary and Conclusions:

> Tom Moody
> Mayor, Columbus, Ohio and Vice President, National
>> League of Cities

Appendix B

DIRECTORY OF SOME EUROPEAN ORGANIZATIONS INVOLVED
IN DEVELOPING RESOURCE RECOVERY SYSTEMS

AB Svenska Fläktfabriken
Fack S-104 60
Stockholm
Sweden

Babcock Krauss-Maffei
Industrieanlagen GmbH
D-8000 München 50
Tannenweg 4
West Germany

Babcock & Wilcox, Ltd.
Grosvenor Gardens House
35-37 Grosvenor Gardens
London SW1W OBS
England

S.p.A. Forni ed Impianti
     Industriali De Bartolomeis
20124 Milano
Via Settembrini 7
Italy

Esmil Habets
St. Anna, Str. 2
Nymegen
Netherlands

Hazemag S. a. r. l.
D-44 Münster
Postfach 3447
West Germany

Imperial Metal Industries (Kynoch)
     Ltd.
P.O. Box 216
Witton, Birmingham B6 7BA
England

Newell Dunford Engineering, Ltd.
Misterton, Doncaster
South Yorkshire DN10 4DN
England

PLM
Fack S-201 10
Malmö
Sweden

RDF, Ltd.
London N17 0DH
England

SOCEA
280, Avenue Napoleon Bonaparte
Boite Postale No. 320
92506 Rueil-Malmaison
France

Soceting
3, Rue Largilliere
75016 Paris
France

Sorain-Cecchini S. p. A.
Via Bruxelles 53-00198
Rome
Italy

Warren Spring Laboratories
P. O. Box 200, Gunnelswood Road
Stevenage, Herts SG1 2BX
England

TNO
Central Technical Laboratory
Appledoorn
Netherlands

Appendix C

ADVANCE LETTER OF INTENT TO BID FOR THE PURCHASE OF RECOVERED PRODUCTS

WHEREAS, the _____ Corporation (hereinafter called the CORPORATION) endorses resource recovery from municipal solid waste as a means toward a cleaner environment and preservation of natural resources, and

WHEREAS, the CORPORATION recognizes the need to develop firm expressions of intent to purchase materials or energy products recovered from waste within known financial parameters as part of the planning process for a new endeavor such as this, and

WHEREAS, _____ (hereinafter called the DEVELOPMENT AGENCY), is evaluating the prospects of substituting resource recovery for its traditional means of solid waste disposal, and

WHEREAS, the DEVELOPMENT AGENCY recognizes the need to establish financial underpinnings for the determination of the economic feasibility of processing up to _____ tons per day of municipal solid waste to produce up to _____ tons per day of _____ (hereinafter known as the PRODUCT) in a form usable and acceptable to the CORPORATION according to the Specifications attached to this *Agreement* and made part hereof,

THEREFORE, in consideration of the fact that the legal authority to sell recovered products may rest upon a requirement to advertise for the purchase of such products, it is mutually agreed between the CORPORATION and the DEVELOPMENT AGENCY that:

I. The CORPORATION, as an expression of its support of the municipal solid waste recovery program, agrees to:

(1) Offer herein a firm commitment to bid for the purchase of the recovered PRODUCT at prices not less than those entered here should the DEVELOPMENT AGENCY be required or decide to effect a competitive procurement, and

(2) Agrees that if public bidding is not necessary and not the course chosen by the DEVELOPMENT AGENCY then the conditions of this Letter of Intent may be considered as a bona fide offer to purchase the recovered product at prices not less than those entered here.

(3) Respond should a bid be required with a bona fide offer to purchase which will include the following:

(a) It will be a firm bid for five (5) years offering an Exchange Price either fixed or related to a commodity quote, and if the Exchange Price is not fixed, it will offer a Floor Price below which the Exchange Price will not fall during the term of the contract.

(b-1) If the Exchange Price to be paid by the CORPORATION is to be fixed dollar amount per unit of product, f.o.b. the recovery facility (or the CORPORATION'S plant - choose one), the bid shall not be less than _____ per ton.

OR

(b-2) If the Exchange Price is to be based on a commodity quote, the monthly Exchange Price shall relate to the quotation at the close of that month for _____ (the same or the appropriate analogous commodity and location) as published in the last issue of the month of _____ (fill in source of quote) using the (midrange or highside, or lowside—choose one) of the quote, f.o.b. the recovery facility (or the CORPORATION'S plant - choose one). If the Exchange Price is to be bid in terms of a percentage of the quoted price, the Exchange Price shall not be bid at less than _____ percentage of appropriate quote as defined above. (Fill in percentage).

(c) If the Exchange Price is not fixed, a Floor Price will be bid which will not be below $_____ per ton f.o.b. (fill in dollar amount) the recovery facility (or CORPORATION'S plant — choose one).

(d) The CORPORATION shall retain the right to reject any material delivered which does not meet Specifications. Such rejection will be at the expense of the resource recovery plant.

(e) The bid will be subject to *force majeure*.

(f) It will be noted the Additional Conditions of the CORPORATION covering general

163

terms and conditions of purchase, acceptance, delivery, arbitration, weights, and
downgrading not explicitly covered in this
Letter of Intent or by reference, will be
negotiated according to good business practices and include such Additional Conditions as are attached to this *Agreement* and
made a part hereof.

(g) This Advance Letter of Intent to bid is null
and void if during the period between its
execution and the actual bid or negotiated
contract the CORPORATION'S plant ceases
operation or ceases use of this or equivalent
grade of recovered PRODUCT. The JURIS-
DICTION shall further recognize that a
clause similar to this shall be incorporated in
the actual bid when made or contract when
signed.

(h) This Advance Letter of Intent may be
assigned by the DEVELOPMENT AGENCY.

II.   The DEVELOPMENT AGENCY agrees:

(1) To see that the recovery plant establishes specification assurance procedures for the recovered PROD-
UCT, using good industrial quality control practiced in recognition of the CORPORATION'S use
technology as practiced in their_____
plant, so as to produce and offer the recovered
PRODUCT for sale in a form and to the required
Specification, usable in the plant with minimum
alterations to present processing technology and
business practices, and

(2) To require, should a contract be effected as a result
of the Advance Letter of Intent, that the PROD-
UCT be delivered to the CORPORATION according to conditions and prices determined herein and
not diverted to the spot market which may on
occasion be higher than the Exchange Price determined by the pricing relationship set forth here or
as modified by the contract.

(3) That should the CORPORATION'S plant, as specified herein, become saturated in its ability to
handle the recovered PRODUCT as a result of
other Letters of Intent issued by the CORPORA-
TION being converted into firm contracts for
delivery and purchase prior to effecting such
arrangements as a result of this commitment, the
provisions of this Advance Letter of Intent become
null and void.

The CORPORATION will communicate to the DEVEL-
OPMENT AGENCY that information about its use technology and business practices which the CORPORATION at
its sole discretion shall consider necessary so as to assure
receipt of the recovered material in form and cleanliness
necessary for use by the CORPORATION. Such communication shall be on a nonconfidential basis, unless otherwise
subject to a subsequent confidentiality agreement.

This Advance Letter of Intent shall become null and void
on _____ unless effected into a contractual relationship
or mutually extended by both the CORPORATION and
DEVELOPMENT AGENCY.

Witnessed by: DEVELOPMENT AGENCY

_____   By: _____

Witnessed by: CORPORATION

_____   By: _____

Source: Technical manual for the Pennsylvania Solid Waste-Resource
Recovery Development Act, 1975 (Commonwealth of Pennsylvania,
Department of Environmental Resources), a publication prepared by
the National Center for Resource Recovery for Pennsylvania's Department of Environmental Resources.

INDEX

165